A STUDENT'S BOOK ON SOILS AND MANURES

A STUDENT'S BOOK
ON
SOILS AND MANURES

BY

E. J. RUSSELL, D.Sc., F.R.S.

DIRECTOR OF THE ROTHAMSTED EXPERIMENTAL STATION
HARPENDEN

THIRD EDITION

CAMBRIDGE
AT THE UNIVERSITY PRESS
1940

CAMBRIDGE
UNIVERSITY PRESS

University Printing House, Cambridge CB2 8BS, United Kingdom

Published in the United States of America by Cambridge University Press, New York

Cambridge University Press is part of the University of Cambridge.

It furthers the University's mission by disseminating knowledge in the pursuit of education, learning and research at the highest international levels of excellence.

www.cambridge.org
Information on this title: www.cambridge.org/9781107654341

First edition 1915
Second edition 1919
Reprinted 1921, 1928
First published 1940
Third edition 1940
First paperback edition 2014

A catalogue record for this publication is available from the British Library

ISBN 978-1-107-65434-1 Paperback

PREFACE TO THE THIRD EDITION

THIS edition has been completely revised and largely rewritten, so as to utilise as fully as possible the newer material and experience available since the last issue. I have had in mind the fact that many of the students using the book would afterwards proceed to work in the Empire overseas and so I have dealt with certain problems such as soil erosion which are more important elsewhere than at home. Throughout I have endeavoured to steer a straight course between the elaboration which most students would find unnecessary, and the over simplification which often leaves the student with a good deal to unlearn if he proceeds to a fuller study of the subject. Further, I have left it to the teacher to decide how much if any formal chemistry should be introduced.

The proofs have been read by Mr H. W. Gardner of the Hertfordshire Institute of Agriculture, and Mr F. Knowles of the East Anglian Institute of Agriculture, both of whom have made many helpful and constructive suggestions based on their wide and successful experience of the needs of farm institute students. To them I wish to tender my best thanks, and also to my colleague, Mr H. V. Garner, who, as always, has helped me considerably in preparing the material.

E. J. R.

ROTHAMSTED EXPERIMENTAL
STATION, HARPENDEN

January 1940

TABLE OF CONTENTS

Appendices

PART I

AN ACCOUNT OF THE SOIL

CHAPTER I

WHAT THE PLANT NEEDS FROM THE SOIL

It is impossible for anyone to know all about any natural object, however simple it may appear. A wheat plant looks at first sight as if it were an easy thing to study, yet in spite of years of work a chemist would have to confess himself unable to give a complete account of the substances it contains, and a botanist would have to admit that much of its structure is unknown to him. And so it is with the soil. Chemists, physicists, geologists, bacteriologists and others have all studied it, but those who have done most would be the first to admit that we really know very little about it, and much still remains to be discovered.

The farmer or the gardener is chiefly interested in soil as the place where his plants grow, and this aspect of the soil, its relation to plant growth, is particularly investigated in agricultural laboratories. Before it can seriously be studied we must first know what the plant requires from the soil: we can then proceed to see how and in what way the soil fulfils these requirements. It is the business of plant physiologists to ascertain plant requirements, and we must therefore start out with the information they have provided.

Six conditions or factors are known to be necessary before the plant will make good growth: the soil must supply a suitable amount of: (1) food, (2) water, and (3) air; (4) it must be at a proper temperature; (5) there must be enough of it to afford adequate root room; (6) it must be free from injurious conditions or pests. What exactly is a suitable amount cannot be stated beforehand but can only be found out by trial; because different plants, and even different varieties of the same plant, have different requirements. Thus an azalea needs all the six conditions and so does a barley plant, but the suitable amount is very different in the two cases. Unfortunately, no way of finding out the suitable amounts has yet been discovered, except actual trial, and this, though it looks straightforward, is really cumbersome and liable to give misleading results.

LIMITING FACTORS

No one of these six conditions can take the place of any other. If a plant is dying for lack of water it will not recover by receiving more food or more air. A proper supply of all the factors must be maintained, and if any one is insufficient or excessive the plant suffers. It is convenient to use a special name for the condition the insufficiency or excess of which is preventing the plant from making better growth, and to speak of it as the "limiting factor". Thus on a dry chalky soil the water supply is often the limiting factor; if more water is got into the soil a bigger crop will be obtained. In a cold summer the temperature is not infrequently the limiting factor; had the days and nights been warmer the plants would have made more growth. On poor soils the food supply is the limiting factor, and addition of more food in the form of manure will increase the crop. The problem of successful management of soil fertility resolves itself into finding out what is the limiting factor and then correcting it as cheaply and completely as possible. This is easy on paper but often difficult in practice.

Fig. 1. Tomatoes growing on a light sand with varying food supply. Pot 47, without manure. Pot 55, one dose of manure. Pot 63, two doses of the same manure.

Where no limiting factor is operating it frequently happens that if one of the necessary factors is increased in amount there will be an increase in crop growth. This is shown in Fig. 1 illustrating three pots of tomatoes growing in the same soil, sown at the same time and

Plot 3 5 6 7 8
6·7 7·8 12·5 17·6 20·1

Mean yield of grain, cwt. per acre

Fig. 2. Effect of increasing dressings of fertilisers on the yield of wheat, Broadbalk, Rothamsted.

Plot 3. No manure.

Plot 5. Manure complete except for one constituent—nitrogen is omitted.

Plot 6. Complete manure containing 43 lb. nitrogen per acre.

Plot 7. Complete manure containing 86 lb. nitrogen per acre.

Plot 8. Complete manure containing 129 lb. nitrogen per acre.

treated alike in every respect except one. The soil is a very light sand; in one pot there has been no addition of plant food; in the second the crop has received a dose of manure, and in the third it has received a larger dose. A similar result is obtained in the field as shown in Fig. 2; the shortest wheat plant is a representative specimen of the crop on the unmanured land; the next plant shows what happens when an almost but not quite complete manure is added, the really essential constituent being left out; the third shows the marked gain when one dose of complete manure is given; next comes the effect of two doses; and the last shows the effect of three doses. In all cases an increase in the amount of plant food has led to an increase in the crop.

Very similar results are obtained when the water supply is varied. In Fig. 3 are shown tomato plants growing in a good soil, sufficiently and equally manured, and under the same favourable conditions of light, temperature, air, etc. All the conditions, excepting one, are the same for all pots: the water supply only varies. When only little water is given the growth is poor in spite of the presence of food and the favourable temperature and light conditions; when more water is added there is better growth; finally with adequate water supply growth is really good.

THE LIMIT OF PLANT GROWTH

But growth will not go on indefinitely. A limit is reached sooner or later beyond which the plant will not make any more growth no matter how much food or water is given. Indeed it is easy to overstep the limit and give too much so that the yield actually suffers.

Pot No. 17 19 21 24

Fig. 3. Tomatoes grown in good soil, all equally manured, but receiving different quantities of water.

Pot 17. No water added.
,, 19. 5 per cent added, and the moisture then kept constant.
,, 21. 10 per cent added ,, ,, ,,
,, 24. $12\frac{1}{2}$ per cent added ,, ,, ,,

Pot No. 47 55 63 72 79

Fig. 4. Tomatoes supplied with increasing doses of manure.

Pot. 47. No manure.

Pots 55 to 79. Increasing dressings of manure. This increases the amount of growth and of fruit up to pot 72 but it depresses yield of fruit in pot 79 where too much is given. The middle pot, 63, is best for fruit.

This has happened in the experiment recorded in Fig. 4. Here, as in Fig. 1, tomatoes are shown growing in soils provided with different amounts of manure. The first and second doses of manure resulted in an increased crop: the third dose caused no further increase: while the fourth actually caused a decrease, the excess of food now acting as an injurious substance. This is well seen also in pots 27 and 36, Fig. 5 (top row). The lodging of wheat through excess of nitrogenous manure is another example.

The limit reached in any particular instance, however, is not necessarily the best growth that can be obtained. It may be set by the insufficiency of water, of temperature, etc. Fig. 5 shows in the upper part a set of tomato plants supplied with successively increasing amounts of manure and 5 per cent of water; in the middle a set supplied with the same amounts of manure and 10 per cent of water; and in the lower part a third set also receiving the same quantities of manure but 12·5 per cent of water—this being as much as the soil would hold. The limit of growth reached in the first case is clearly due to a deficiency of water, for it is raised considerably when more water is added. But a still further increase in the supply of water does not lead to more growth, the limit being now set by something else. It is possible that by increasing the temperature or the root room we could get more growth out of this last series, but the process comes to an end before long and the final limit is set by the sheer inability of the plant to grow any bigger. If larger crops are wanted it becomes necessary to try some bigger yielding variety, i.e. some plant with more power of growth.

Pot No. 3 11 20 27 36

5 per cent water.

Pot No. 5 13 21 30 38

10 per cent water.

Pot No. 7 15 24 32 39

12½ per cent water.

Fig. 5. Tomatoes grown in soil receiving successively increasing doses of manure in pots passing from left to right. Pots 3, 5, 7, no manure; pots 36, 38, 39, ten doses manure.

Top row: moisture maintained at 5 per cent.
Middle row: ,, ,, 10 ,,
Bottom row: ,, ,, 12½ ,,

All these results are shown in the curves of Fig. 6. But there is something more than actual weight. The student who carries out the experiment will observe that some of the plants differ very much in appearance and agricultural or horticultural value even when their weights are not unlike. Between pots 3 and 7 (Fig. 5), for instance, there are great differences in appearance and habit of growth. Pot 3 (5 per cent of water and

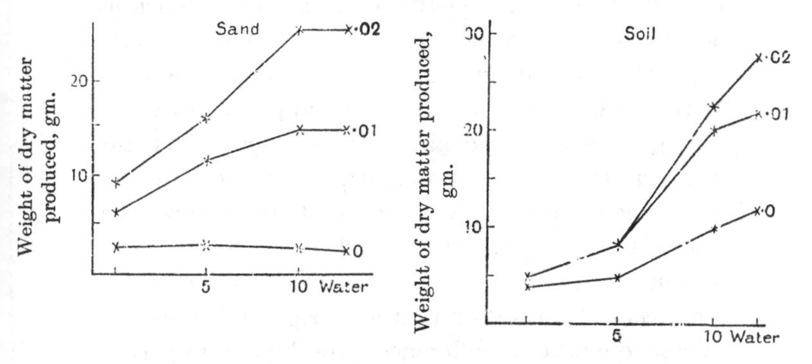

Fig. 6. Curves showing weights of crop produced with varying supplies of water and 0, 0·01 and 0·02 g. of nitrate of soda per pot.

no nitrate) contains sturdy plants capable of great development if transplanted into more favourable conditions, while pot 7 ($12\frac{1}{2}$ per cent water and no nitrate) contains "leggy" plants that would never be of any value. Similarly the wetness of the soil affects the root development: in a dry soil there is more root than in a wet one: von Seelhorst showed that barley growing in a soil watered only to half its full water-holding capacity produced twice as much root as when the water was maintained at three-quarters the full capacity.

Another important result of differences in moisture

content is the effect on ripening of cereal crops. In very dry conditions cereals ripen early and the grain may be small and shrivelled: in good moist conditions ripening is much better, especially if the water supply is reduced at the end: but in continuous wet conditions it is delayed: less grain is formed and more straw. This is well shown by growing barley in place of tomatoes in the experiment shown in Fig. 5. The experiment was made many years ago by a German agriculturist Hellriegel, and the results are plotted in Fig. 7. With no water supply there was no growth: with low supply there was a little but not much; as the water supply increased the plant growth increased also. But the grain increased more than the straw: the weight per corn went up also. Then came the turning point: beyond about 40 per cent of saturation less grain was formed. The straw, however, still continued to increase, but when the water exceeded 60 per cent of saturation that also began to suffer.

These qualitative differences are highly important from the practical point of view but they are much more difficult to investigate than mere changes in weight.

From these and similar experiments we may deduce three general principles of the highest importance in the study of soil fertility:

(1) Six separate soil factors are necessary for the successful growth of the plant: there must be an adequate supply of food, water, air, a suitable temperature, sufficient root room and an absence of harmful substances. If any of these conditions is not complied with the plant fails to grow well: the lacking condition is called the *limiting factor*, and it must be supplied before further growth takes place.

(2) By increasing the supply of any of the factors necessary for the plant but present in insufficient

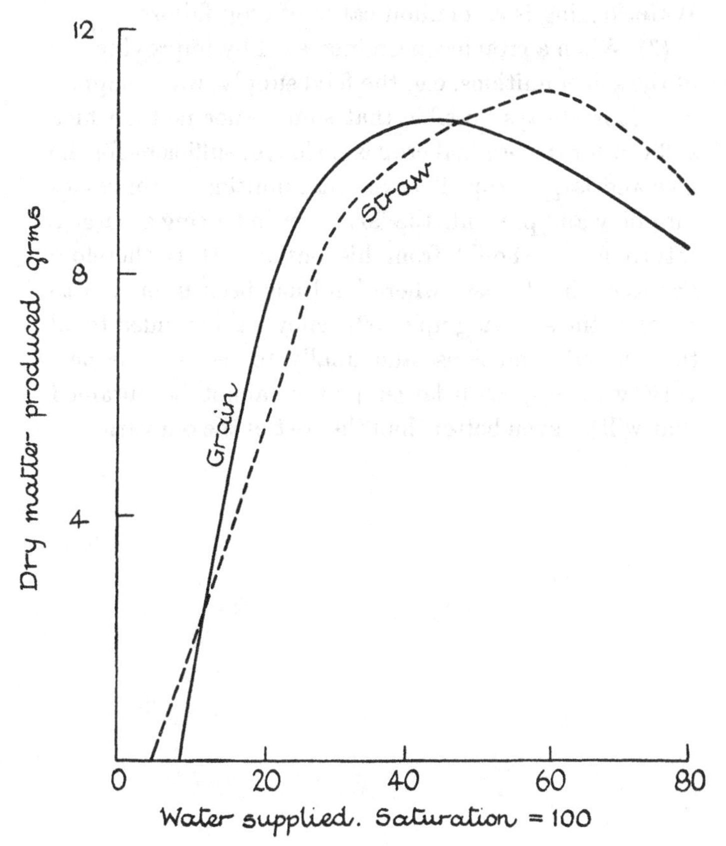

Fig. 7. The relation between water supply and plant growth. With increasing water supply the yield of grain rapidly increases up to a maximum and then falls off. The straw increases even at higher water content but that too falls off after a time.

quantity an increase in growth is obtained. But a limit is sooner or later reached beyond which further growth will not take place. Additional increases in the food, water supply, temperature, etc., may do positive harm. Waterlogging is a common cause of crop failure.

(3) When a crop has been increased by improving one of the soil conditions, e.g. the food supply, water supply, etc., it is always possible that some other factor which sufficed for the original crop is no longer sufficient for the new and larger crop. Thus another limiting factor comes into play and prevents the farmer from getting as large a return as he should from his outlay. It is therefore necessary in all cases where land has been improved to see that the screwing up of efficiency has extended to all the six soil conditions, and finally to see if some new variety of crop with larger power cannot be obtained that will do even better than the best of the old varieties.

CHAPTER II

THE COMPOSITION OF THE SOIL

THE reader must often have noticed in walking along a lane after a heavy rainfall that the water streaming down a bank has washed away the soil in a somewhat uneven manner, leaving behind the grit and small stones but carrying away the rest. In following the course of such a streamlet one observes that at certain points a smooth cake is formed which cracks as soon as it begins to dry, and is much more sticky and clay-like than the original soil. Closer observation shows that the original soil has been separated into various constituents by the running water, the heavier coarser particles being left behind while the finer lighter particles are carried on.

This effect of a flowing stream has suggested a method for analysing soil that in its later forms has proved very useful and has been largely adopted by soil investigators. It consists of allowing a stream of water to flow over the soil and to sort out the particles according to their degree of fineness. One form of apparatus for doing this, designed by Nöbel, is illustrated in Fig. 8. 25 gm. (about 1 oz.) of soil are put into the smallest of the pear-shaped vessels *A*, and water is run in from *W*. As the vessels are of different diameters the water flows through them at different rates, going most rapidly through the narrowest and most slowly through the widest, *D*. When it runs rapidly it carries away the fine and intermediate particles leaving only the coarsest: as it goes more and more slowly it deposits finer and finer particles. Hence after a time the soil put into the apparatus has become sorted out into grades, the coarsest particles only remaining in the smallest vessel *A*, while the other portions of successively finer particles are distributed over the larger vessels *B*, *C*, and *D*, till finally the smallest particles of all get washed out into the large vessel *E*.

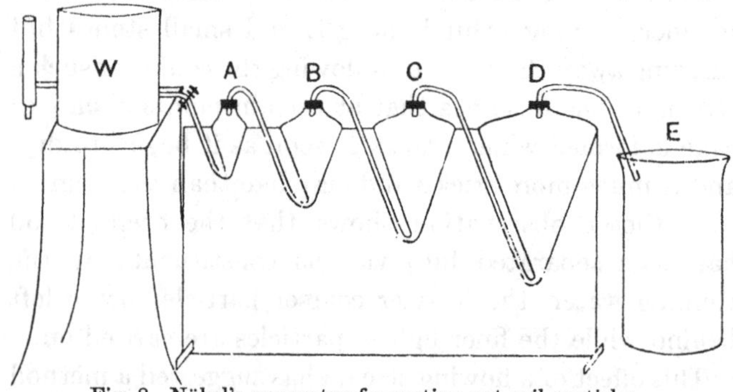

Fig. 8. Nöbel's apparatus for sorting out soil particles.

Instead of allowing the water to flow over the soil the separation may equally well be brought about by allowing the soil to fall through water. A simple apparatus devised for the purpose by J. Alan Murray is shown in Fig. 9. A long glass tube about 50 in. long and 1 in. wide is fitted by means of a wide piece of rubber tube to a 200 c.c. conical flask with a neck 1 in. wide containing 5–10 gm. of soil. The flask is half-filled with water and vigorously shaken so as to break up the soil; then it is almost completely filled with water and attached to the long tube. The whole apparatus is now filled up with water and inverted in a vessel of water. Instantly the soil begins to tumble through the water, but some of it falls more quickly than the rest. The large coarse particles reach the bottom of the tube very quickly and form a little layer there, or, if the tube is left open, they can be collected in a small dish. Next come the small but still coarse particles. After these the fine particles begin to come down, and at the end the finest of all settle as a light mud. The experiment is even more striking if it is carried out with a mixture of various grades of clean sand.

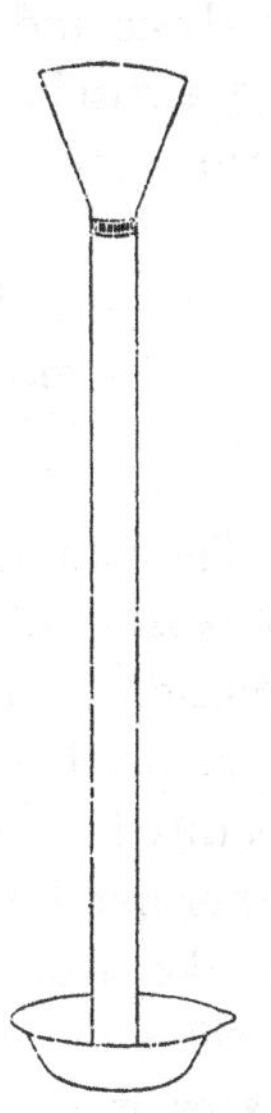

Fig. 9. Murray's apparatus for sorting out soil particles.

More refined methods are in use in analytical laboratories. The lumps of soil are first broken down by a wooden pestle and then by treatment with certain chemical reagents. Next the soil is passed through a sieve of known dimensions which sorts out the particles of a certain size. Then it is shaken up with water and allowed to settle for certain times. It can be shown mathematically that the speed with which a particle sinks through the column of water is proportional to the area of its cross-section, i.e. to the square of its radius, hence the method enables us to grade the particles according to their size. Numerous investigations have brought out the remarkable fact that the soil contains many particles as small as $\frac{1}{12500}$ in. (0·002 mm.) or less in diameter. The fine earth used in laboratory tests is the material of which the particles are less than $\frac{1}{12}$ in. (2 mm.) in diameter.

In soils of temperate climates no natural division

usually occurs between the various fractions; the particles merge by imperceptible gradations from the very coarsest to the very finest. In tropical soils, however, the middle fractions are sometimes almost absent. It is convenient to make divisions for the purpose of analysis and investigation, but except for the clay they are entirely arbitrary. In this country the following grades are now adopted:[1]

	Diameter of particles
Gravel	Above 2 mm.
Coarse sand	Between 2 and 0·2 mm.
Fine sand	Between 0·2 and 0·02 mm.
Silt	Between 0·02 and 0·002 mm.
Clay	Below 0·002 mm.

The material sorted out in the above experiments can be used to discover some of the properties of the various fractions. The coarsest material on examination is found to be hard and gritty, to dry quickly and to separate out readily into individual grains. The finest material, on the other hand, is soft and smooth, it dries slowly and forms a cake which cracks into little flakes that curl up in a curious manner. If one of these flakes is dropped into water it falls to the bottom in one piece, but if it is rubbed between the fingers under water it breaks up into particles so minute that they do not settle but make the water turbid.

The question at once arises: Why are the particles so different in size? Why are some so small and others so large? An obvious answer is that the large particles are perpetually breaking up into little ones and that the silt represents a sort of half-way stage between gravel and clay. This, however, is not the whole explanation. The

[1] The older grades were rather different.

clay contains substances that are not in the sand, as is shown by the following experiment. Put into one test-tube 1 gm. of the sand and into another 1 gm. of the clay: add 20 c.c. of strong hydrochloric acid to each, plunge the test-tubes into a vessel of boiling water and leave for an hour. Hydrochloric acid is a potent solvent, and dissolves material that is not highly resistant. At the end of an hour the clay is seen to yield a markedly coloured solution while the sand only gives a slightly yellow solution: dilute and filter these and add ammonia to each until the liquid turns red litmus blue: the solution from the sand gives only a slight precipitate while that from the clay gives a much denser one. Thus we conclude that sand is much more resistant to the attack of acids than clay. The same result is obtained when the sand and the clay are exposed to the weathering agencies: the sand resists more than the clay and therefore is less completely broken down.

We must now proceed a stage further and try to discover how the particles got there and what their history has been.

THE ORIGIN OF THE SOIL PARTICLES

The soil particles have originally been derived from the rocks, and their present state is the outcome partly of the nature of the rock from which they arose and partly of the circumstances through which they have passed. The original rock was formed from molten material and is therefore called "igneous", it might be either acidic (e.g. granite) or basic (e.g. basalt) according as it was rich or poor in silica. In Great Britain, however, soils of igneous origin are not agriculturally important:

most of our soils have had a much more varied history. The original rock gradually crumbled by alternate warming and cooling and by the action of water or ice; some decomposition took place; the materials formed were carried by percolating water, wind, rivers or glaciers for a greater or less distance and ultimately found their way to the sea and there they were deposited. In course of time the pressure of the great accumulation of material caused some of it to be converted again into rock and, when the sea-floor was uplifted to form dry land, this new rock thus exposed went through the same processes of disintegration, and again the particles were exposed to air, to water and to ice. Sometimes they remained where they were, or were carried only short distances: sometimes they were carried away a great distance. In many districts, as in Central and Eastern Europe, parts of Asia and of the Middle West of the United States (e.g. in Nebraska), wind was the transporting agent, and the soils thus formed, known as loess soils, are remarkable for the narrow range of variation in size of their particles, the wind being able to carry particles of certain dimensions only. Over much of England north of the Thames, and the northern parts of the United States, glaciers carried the particles to their present position, grinding them sometimes almost to impalpable powder; these are the boulder clays, the glacial sands and gravels, etc. Elsewhere flowing water was the transporting agent; these are called alluvial soils. The particles as we find them to-day are largely the result of their past history; indeed, in many cases the properties of the soil were mainly settled in geological ages far remote from our own. The red Triassic soils of the west of England were formed under continental conditions with much wind-drifted material; they are quite distinct in character from the poor clays of the older Coal Measures or the grey soils of the succeeding Lias. These all differ considerably from the Oxford Clay and from the Weald Clay.

SOIL CONSTITUENTS.

CALCIUM CARBONATE, CHALK, LIMESTONE

Calcium carbonate is one of the simplest and most important of the soil constituents, though it is sometimes present only in very small amounts. It occurs in several forms which, though chemically identical, differ in other ways. As chalk, it covers a large area of the eastern half of England, including portions of the counties eastwards of the line joining Lincolnshire and Wiltshire. Limestone occurs extensively in the eastern counties, Lincolnshire, Northamptonshire, Oxford, Gloucestershire and Wiltshire as an oolite which impresses a definite character on the soil. Both chalk and oolite are the remains of minute sea animals, as may be seen by examination under a microscope.

Calcium carbonate occurs also as dolomitic and carboniferous limestone in the north and west, but here it is hard and has much less effect in determining soil properties.

Small quantities of calcium carbonate are found in many soils. In some cases it has been brought in from the chalk or limestone by water or by glaciers in times gone by, as in the chalky boulder clay of the eastern counties. Some is formed also during the weathering of rocks, and some from the decomposition of plant and animal remains. In some districts a good deal of chalk has been added to the soil by farmers in the past: some of the fields in Hertfordshire still contain as much as 1 or 2 per cent put on as top dressings a century or more ago.

One of its most familiar properties is that it changes to quicklime when strongly heated in a lime kiln: the process is called "lime burning", though the name is not a good one because neither chalk nor lime can burn. Two tons of chalk produce about one ton of lime. The loss of weight is due to the giving off of a gas, carbon dioxide, which escapes into the air with the waste gases from the fuel.

When the experiment is done carefully on a small scale in the laboratory it is found that 100 parts by weight of pure calcium carbonate, after removal of all impurities, gives rise to 56 parts by weight of lime and 44 parts by weight of carbon dioxide. This relationship is very important, for it shows how the carbonate is built up: it may be expressed thus:

Calcium carbonate =	lime or calcium oxide +	carbon dioxide
100	56	44 parts by weight

Carbon dioxide is of vital importance in agriculture. It is breathed out by human beings and by animals, and it would put an end to our lives if it accumulated. Fortunately for us it is absorbed by living plants, supplying them with the carbon which forms about half of their dry weight. It dissolves in water, forming an acid called carbonic acid; this is weak and unstable but it produces some remarkable effects in nature chiefly because of its solvent power. One million parts of pure water dissolve only 13 parts by weight of calcium carbonate at 10° C., but one million parts of water saturated with carbon dioxide dissolve about 900 parts. The explanation is that the calcium carbonate is changed by the carbonic acid into the more soluble calcium bicarbonate. Rain water contains some carbonic acid and soil water still more, so that in chalk or limestone districts the spring and well waters are rich in calcium bicarbonate. The carbon dioxide is driven off, however, when the water is boiled, and the calcium carbonate is then re-formed and thrown out of solution: this is how the "fur" in kettles, or the "scale" in boilers, is formed. Sometimes it even deposits on standing, forming a sediment in the vessel or a crust on any object lying in the water. Knaresborough in Yorkshire has a well-known spring where this deposition takes place. In some districts so much carbonate is dissolved out from the soil that great gorges and caverns are formed, as at Cheddar in Somerset, in the Peak district of Derbyshire and the Ingelborough region of the West Riding.

The equation set out above shows that calcium carbonate is built up of two oxides: calcium oxide formed by the union of the metal calcium with oxygen, and carbon dioxide formed by the union of the non-metal carbon, the chief constituent of charcoal, with oxygen. The calcium is called the base; it is characteristic of bases that they can react with acids to produce salts. Thus calcium carbonate is a salt.

The acid part of the salt can be displaced by another acid which is stronger or present in much greater quantity. Instead of removing the carbonic acid by heat, as is done in lime burning, it can equally well be done by means of a stronger acid such as hydrochloric acid. The reaction is

Calcium carbonate + hydrochloric acid = calcium chloride + carbonic acid

but the carbonic acid, being unstable, rapidly breaks down to form water and carbon dioxide which comes off with vigorous effervescence.

This explains why calcium carbonate is so valuable on acid soils, as we shall see later; it interacts with the harmful soil acids, forming calcium salts and carbonic acid, all of which are harmless.

This decomposition of calcium carbonate is the basis of a simple qualitative test for detecting its presence in soils. Put about 5–10 gm. of soil (about $\frac{1}{4}$ oz.) into a cup and cover with water, stir with a piece of wood or glass rod till all bubbles have gone: then add hydrochloric acid drop by drop. Even if only 0·05 per cent of calcium carbonate is present it can still be detected by the effervescence. The test has been elaborated by chemists and made quantitative, based on the fact already set out above that 44 parts by weight of carbon dioxide correspond to 100 parts by weight of calcium carbonate.

Just as the acidic part of a salt is replaceable by another acid, so the basic part is replaceable by another base. A solution of copper sulphate is blue, but if it is used like ink in writing on a zinc label it makes a permanent black mark. Some of the zinc and copper have changed places, copper is deposited as a metal on the zinc where it appears black, and zinc has gone into solution as a colourless sulphate. Some important exchanges of bases are dealt with later.

SILICA

One of the commonest mineral substances on the earth is silicon dioxide or silica, the chief constituent of quartz, flint, and most sands. In these forms it is so hard that it can only be powdered with difficulty; it is also only very slightly soluble in water. The sand on the seashore affords sufficient illustration of its properties: in spite of the persistent hammering of the waves, the washing of the sea and the rain, and the exposure to all sorts of weather, it undergoes no perceptible change in any period within the memory of an individual; the sand may be carried away but it does not appreciably dissolve or break down under the influence of these agencies. The immediate origin of a sandy soil is commonly a sandstone rock which is itself composed of grains of sand united by some kind of cementing material. When the rock was exposed to the action of the weather the cement was washed away and then the whole structure fell to pieces, grains of sand having little or no power of holding together by themselves.

THE SILICATES

As its chemical name implies, silica is the oxide of the non-metal silicon; it is analogous to carbon dioxide and it unites with water to form a weak rather unstable acid called silicic acid which combines with bases forming silicates. Water-glass is a familiar example: it is sodium silicate. Ordinary glass is more complex, containing silicates of sodium and calcium in a special physical condition. The rocks from which soils are formed consist largely of silicates. Some are relatively simple and something is known about their make-up; orthoclase felspar, the pink crystals occurring in some of the granites, is a silicate of potassium and aluminium. Others are much more complex and may include iron, aluminium, calcium, magnesium, sodium, potassium, etc., some are almost as resistant as sand to the action of water and weather so that they remain as relatively coarse particles and behave agriculturally like sand. Others, however, are more easily acted upon, with two important results. Instead of being inert, like sand, they decompose under the influence of carbonic and other acids in the soil, forming a variety of new products, some of which are reactive, i.e. they act upon various substances brought in contact with them; some are like jelly, and some remain as particles but are extraordinarily small and form the clay, the particles which tail off from 0·002 mm. ($\frac{1}{12500}$ in.) in diameter to much smaller dimensions.

PROPERTIES OF SAND, SILT AND CLAY

The most striking property of sand is its great resistance to the action of water and weather; this gives rise to consequences of great agricultural importance. Sand gives up little or nothing to plants and hence is in no sense a plant food; indeed, plants quickly starve in it. Its particles vary, as we have seen, from 2 mm. ($\frac{1}{12}$ in.) to 0·02 mm. ($\frac{1}{1250}$ in.) in diameter. Even the edges do not easily wear away and the particles remain irregular in outline. Their large size and irregular shape prevent them from packing very closely, and large pore spaces are left in between. Consequently air gets in very easily, water rapidly flows through, and the sand speedily dries, at any rate near the surface.

CLAY

Clay differs very widely from sand in its properties. It is very sticky and belongs to the class of substances called "colloids", a word which means "like glue". It is such a dominating substance that even a small quantity impresses its properties on a large bulk of silt or sand, and a soil containing 20 per cent of clay easily becomes quite sticky and hard to cultivate.

The small size of the clay particles enables them to pack closely together, leaving only minute pores or passages along which water can pass. The total volume of pore space is more variable than in sandy soils, and in field conditions may range from 30 to 50 per cent, while in sandy soils the range is less, and may be from 30 to 40 per cent. The total pore space may thus be about the same or even greater in a clay than in a sand, but the individual pores in the clay are much smaller though much more numerous. Clay soils are commonly called "heavy" and sandy soils "light": these words, however, refer only to ease of cultivation and not to their relative weights. This can be shown by drying and powdering a very "heavy" clay and a very light sandy soil, and pouring 50 gm. of each into separate 50 c.c. cylinders. The clay just about fills the 50 c.c., while the sand occupies only about 30 c.c. Then pour into separate 100 c.c. cylinders each containing 50 c.c. of water. Both the clay and the sand cause the water to rise about 20 c.c. This shows that the true volume of each is about the same, and as their weight is the same it follows that their density or weight per unit volume is also the same: i.e. about $\frac{50}{20} = 2 \cdot 5$. When the experiment is done more carefully it frequently happens that the density of the sand is rather higher than that of the clay so that it is really somewhat heavier and not lighter. As the clay occupied so much more space than the sand its pore space was evidently much greater: as the result of the drying and powdering the difference between the clay and the sand has become more marked than in the natural state.

In consequence of the smallness of its pores clay impedes the movement of water in the soil. It absorbs a good deal of water and holds some of it so firmly that plants cannot extract it and they may wilt in spite of the fact that a sufficient amount is there for their needs, but the clay is holding it too tightly for them to obtain. Excess of clay interferes too much with the water movements, parching the soil in dry seasons, even though the permanent water level is near the surface, but making it waterlogged in wet weather, thus impeding the movement of air to the roots and lowering the temperature of the soil.

The adhesive properties of clay cause the soil particles to bind together into the aggregates on which "tilth" depends; soil without clay would be very like a sand heap. Here also, however, excess of clay does harm by limiting the range of moisture content over which the soil can be cultivated; too much moisture makes the soil so adhesive that it sticks to the tillage implements and retards their movements; it also tends to form large clods unfavourable to vegetation. These effects are intensified in wet weather; the soil becomes sticky and must not be worked or the tilth is injured for a long time.

Another effect of a large amount of clay is to make the soil shrink very much on drying, so that large cracks appear in the fields in summer time. The swelling of the clay in wetting may be harmful.[1]

Unlike the sand, the clay is chemically active: it reacts with salts as if it were itself a salt. Indeed it may be regarded as a kind of very complex salt, the acid part being a complex alumino-silicic acid, while the basic part includes aluminium, calcium, magnesium, potassium, sodium and hydrogen. Just as there are different salts of the same general type, so there are different clays: that derived from the Gault, for example, has the typical clay properties much more strongly marked than that derived from the Trias formation. When the clay is put in contact with a solution of a salt an exchange of bases takes place, the clay taking up some of the base from the added salt and in exchange giving up some of its calcium or other bases. This property is of great practical importance because it enables soluble salts to be used as fertiliser; ammonium sulphate, for example, is soluble, but the clay exchanges some of its calcium for the ammonium which it then holds in an insoluble form. A similar reaction takes place between clay and potassium sulphate: this salt is soluble, but by exchange the potassium is fixed or precipitated on the clay.

CALCIUM-SODIUM EXCHANGE: WATER SOFTENING

Water softening is another example of base exchange. The calcium bicarbonate present in many spring and well waters interacts with soap (a sodium or potassium

[1] When the Gezira (Sudan) was first irrigated, the swelling of the clay was so great as to force the masonry of the regulators out of position, sometimes as much as 10 cm., thereby causing considerable trouble.

salt of a complex fatty acid) to form a calcium soap which is a hard curd and useless for washing purposes. So the water is called "hard". The process of water softening consists in replacing the calcium by sodium. The water is allowed to pass through a layer of granules of an insoluble sodium aluminium silicate; this reacts with the calcium carbonate in the water to form insoluble calcium aluminium silicate and sodium carbonate which goes into solution. The water has not lost its saline matter but has replaced calcium by sodium which neither wastes soap nor scales boilers. Periodically the calcium aluminium silicate needs "regeneration", i.e. reconversion to the original sodium compound: this is done by allowing a strong solution of salt to pass through the cylinder and the water that passes out is run to waste. Salt is sodium chloride. The changes that take place are:

(1) During water softening:

sodium aluminium silicate + calcium bicarbonate (in hard water)
= calcium aluminium silicate + sodium bicarbonate. (in soft water)

(2) During regeneration:

calcium aluminium silicate + sodium chloride
= sodium aluminium silicate + calcium chloride.

When calcium chloride ceases to appear the regeneration is complete.

The process can be demonstrated by putting a layer of the calcium aluminium silicate (e.g. permutit, obtainable from a chemist) on a piece of glass-wool in a funnel and pouring some hard water through it. Shake up a measured quantity of soap solution with 50 c.c. of the original water and also with 50 c.c. of the treated water and compare the difference in lathering power.

The bases held by the clay, and capable of being exchanged in this way for other bases, are called "exchangeable bases".

In the clay of a fertile soil the dominant exchangeable base is calcium, which may form 80 per cent or more of all the bases held by the clay. In dry regions of the world where inland seas have evaporated, or where for other reasons much sodium or magnesium salt has got into the soil, one of these bases may become more important and give quite a different character to the clay. A sodium clay is much more sticky than a calcium clay, and it tends also to decompose, forming a certain amount of sodium carbonate which makes the soil alkaline and is very harmful to plants. This interaction of common salt and clay to form a sodium clay explains the injury done to agricultural land when it becomes flooded with sea water. The usual sequence is that the soil is at first porous but it becomes more and more sticky as the salt is washed out; finally, it is impervious to air and water when wet and very hard when dry. The calcium clay has become converted into a sodium clay. The remedy is to regenerate the calcium clay: this is done by assuring good drainage to allow the sodium to be removed, then cultivating and cropping to ensure production of carbonic acid in the soil, and if necessary adding calcium carbonate to supply the calcium required. Detailed studies have been made by D. J. Hissink in Holland in connection with the reclamation of the Zuyder Zee, one of the greatest agricultural achievements in Europe.

In Great Britain and other cool or temperate regions the calcium tends to be displaced by hydrogen and then the clay becomes acid, which again is harmful to plants though in a different way.

An important part of the management of a clay soil consists in maintaining a preponderance of calcium in the clay; dressings of chalk and lime have always been known as good practice on clay soils.

CRUMB FORMATION AND TILTH

Calcium clay can form crumbs which remain stable in presence of water, and which help to build up the soil into the granular structure associated with good "tilth". Sodium clay, on the other hand, does not do this anything like so well. But calcium clay can also go into a state in which the crumbs break down and the clay becomes more sticky again.

The joining up of clay particles to form flocs can easily be demonstrated. Stir up some clay in rain or distilled water, pour off the turbid liquid and divide it into three equal parts. To one add 5–10 c.c. of lime water; to another the same quantity of dilute ammonia solution; leave the third alone. Flocculation is seen to take place rapidly under the influence of lime; the untreated portion settles much more slowly; ammonia almost entirely prevents settling.

Other substances also cause flocculation and the phenomena vary rather widely according to the conditions of the experiment. These flocs, however, do not seem to be the same as the soil crumbs, though there is some superficial similarity.

The effect of lime on drainage can be shown by

putting a layer of clay supported on a perforated disk into each of three funnels: sprinkle lime on one; pour 10 c.c. of dilute ammonia solution on to another. Then pour water on to all three so that it stands at the same height in each funnel: leave for a time. Percolation begins first on the limed clay: next on the untreated clay; but proceeds only slowly if at all on the clay treated with ammonia.

Practical men have long since learned that the crumbly state is good for plants while the sticky state is not, and they have also discovered how to change one into another. Addition of lime, chalk or limestone causes the change to take place rapidly: organic matter (such as farmyard manure, or green crops ploughed in), frost and good cultivation also have the same effect.

SILT

Between the inert sand particles and the reactive clay particles there comes an intermediate grade (0·02–0·002 mm. diameter) differing somewhat from either. As the particles are smaller than those of sand they pack together with smaller pore spaces which retard the movements both of air and water, though the total pore space is not necessarily diminished. Soils containing much silt are difficult to drain: the part just above the pipes gives up its water readily, while the rest of the soil a little farther away does not, but remains wet in spite of the drainage system.

Silt differs from clay in that it is not flocculated and rendered less sticky by the addition of lime, or by frost or cultivation. Thus if a soil contains sufficient silt its stickiness and heaviness cannot usually be remedied by liming though it may be by large dressings of farmyard manure. Such soils occur on the Boulder Clays, the Lower Wealden Beds in Sussex, and elsewhere, and they are always a source of trouble. The simplest plan is to leave them in grass, but even this device is not entirely satisfactory.

MINERAL PLANT FOOD: PHOSPHATES AND POTASH

The constituents dealt with in the preceding paragraphs—the sand, silt, clay and the calcium carbonate—compose almost the whole of the mineral part of the soil. But although the balance is only small in amount it is of great importance to the plant, for it contains an essential article of plant food—calcium phosphate. This substance arose in the first instance from the rocks, but often the material in our soils has already done duty in past ages, and has helped to build up the skeleton of some organism, on the death of which it has again returned to the soil to do duty once more. It is readily detected by igniting a few grams of soil in a basin, then boiling with strong nitric acid, diluting, filtering, and adding to the filtrate a solution of ammonium molybdate. A yellow precipitate slowly comes down containing the phosphoric acid extracted by the acid. Potassium, another essential plant food, occurs in the complex silicates of the soil and some of it appears in the hydrochloric acid extract.

The red or yellow colour of the solution is due in part to the iron present. On neutralising with ammonia a dense red precipitate containing iron and aluminium oxides comes down and can be filtered off: the presence of iron can then be confirmed by the beautiful blue precipitate obtained when the red material is dissolved in a little hydrochloric acid and treated with potassium ferrocyanide solution, or by the very deep red colour obtained when some of the hydrochloric acid solution is almost neutralised with ammonia and then treated with potassium thiocyanate solution.

Other important elements present in smaller quantities are described in Chapter XI.

The phosphorus and potassium compounds are among the most important constituents of the soil from the farmer's point of view. Yet they do not form any very great proportion of the whole, and even in a fertile soil there is often not more than 3 or 4 lb. of either in 1 ton of soil, whilst the amount that plants can get hold of may only be a few ounces.[1] The plant, however, does not need a great deal; 1 ton of mangolds only contains some 10 lb. of potash and 1½ lb. of phosphoric[2] acid, so that the quantities present are not as inadequate as they appear.

SOIL ANALYSIS

It has become customary to talk of the "nitrogen", "phosphoric acid", "potash", "lime", etc., in the soil, but the student must at the outset realise that these do not exist as such in the soil. The nitrogen meant is not nitrogen as it occurs in the air, and which is better spoken of as gaseous nitrogen: it is nitrogen combined with other substances. "Phosphoric acid" does not occur in the soil, but only its compounds, the phosphates;

[1] An acre of soil to a depth of 9 in. weighs 1000 to 1500 tons. Chemists distinguish between the "available" and the "unavailable" plant food.

[2] In accordance with British custom these amounts are stated as the oxides K_2O and P_2O_5.

"potash" and "lime" do not occur, but only potassium and calcium salts. The amounts of calcium carbonate, of potassium, phosphorus and other elements present in soil can be estimated in the laboratory. The process is called "soil analysis"; it was at one time thought to be relatively easy, and to be a safe and simple guide to the manuring of crops. This is not so: soil analysis presents some complex problems and must be linked up with field experiments and where possible with a soil survey.

We have now come to the end of the important mineral constituents of the soil. When such a soil is supplied with water, is properly aerated, and receives a sufficient amount of heat from the sun, it speedily becomes the abode of many plants, insects, worms and other small animals and microscopic living things called generally micro-organisms. As these die their remains mingle with the soil, and so a fresh constituent appears, known as organic matter, which has the distinguishing characteristic that it got there through the agency of living organisms and has the chemical distinction of being easily and completely burnt away. The presence of this organic matter is readily shown by heating some soil on a tin lid or in a crucible; the soil blackens or chars, then little sparkles of fire can be seen, and finally all the combustible part smoulders away leaving only the mineral constituents. The organic matter is so important that it must be dealt with in a separate chapter by itself.

AIR AND WATER IN THE SOIL

The mineral and organic constituents form only some 30 to 50 per cent of the volume of the soil as it occurs in the field: the rest is pore space filled with air and water which are of vital importance to the roots of the plants and to the other living things in the soil. The air somewhat resembles ordinary atmospheric air in composition, but it contains more carbon dioxide and more water vapour; typical samples contain:

	Oxygen per cent by volume	Nitrogen per cent by volume	Carbon dioxide per cent by volume
Atmospheric air	20·95	79·02	0·03
Soil air, arable	20·5	79·2	0·3
Soil air, pasture	18·2	80·2	1·6

When the soil becomes waterlogged, however, the percentage of oxygen may fall very considerably so that insufficient is left for the organisms to carry on their usual functions. Undesirable changes may then set in (p. 117). The presence of more carbon dioxide in the soil air than in the atmosphere is readily demonstrated by driving a ½ in. gas pipe to a depth of 6 in. in the soil and connecting it with a test-tube containing 20 c.c. of baryta water and attached to an aspirator. A similar tube also containing 20 c.c. of baryta water but open to the air is attached to the same aspirator. Set the aspirator working and arrange the connections so that bubbles pass at the same rate through the two lots of baryta water. The one connected with the soil speedily becomes turbid, indicating the presence of carbon dioxide; the other, open to the air, however, only shows turbidity later on (Fig. 10). The turbidity is due to the white insoluble barium carbonate formed by the union of the carbon dioxide with the barium hydroxide in the baryta water; an action similar to the formation of calcium carbonate from carbon dioxide and lime.

The water is held by physical forces in the pores and

the proportion of the pore space filled in an uncropped soil depends on the rainfall, the evaporation and the drainage. The physical forces concerned are associated with the surface of the soil particles. Clay has about the

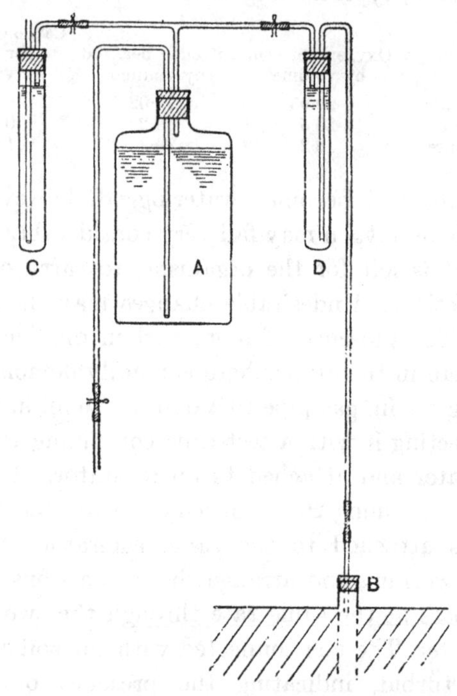

Fig. 10. Apparatus for demonstrating the presence of CO_2 in soil air. *A*, aspirator; *B*, $\frac{1}{2}$ in. gas pipe driven into soil; *C*, tube of baryta water open to air; *D*, tube of baryta water connected to soil.

same density as sand (p. 26) but its particles are much smaller and therefore more numerous than for the same weight of sand; its total surface is consequently greater and so it has more power of retaining water (p. 111).

When more water reaches the soil than can be held by these physical forces the excess drains away, moving downwards unless some obstacle prevents it. On sloping land where the subsoil is rather impermeable some of the drainage water follows the contour and appears at the foot of the slope; this explains the moist condition and the high water table of the valleys. If there is no way out the water simply remains in the soil, waterlogging it. Apart from this downward movement there is very little other water movement in the soil that is of practical consequence. It used to be thought that water moves upwards and sidewards by capillarity. Movement of this kind takes place from a free water surface into dry soil, but not from moist soil; this can be shown by immersing a tube of dry soil in a beaker of water, when the water rises, but nothing visible happens if the tube is placed in a beaker of moist soil.

Plants take out a great deal of water from the soil especially when they are growing vigorously: indeed, the only effective way of drying the lower layers of the soil, apart from drainage, is by transpiration through plants. In the Rothamsted measurements the amount of water in sandy soils was generally found to be about 9 per cent, in loams about 12 per cent, and in clays about 27 per cent by weight; a better idea, however, is furnished by taking the proportions by volume, which vary from 20 to 40 per cent. Figures for some of the Rothamsted soils are given in Table I.

The water is not pure but contains various salts in solution, the most important of which are nitrates and bicarbonates (pp. 21, 48).

Table I. *Volumes of pore space, air and water in soil*

	Vol. occupied in natural state by		Volume of water		Volume of air	
	Solid matter	Air and water (pore-space)	In normal moist state	After period of drought	In normal moist state	After period of drought
Poor heavy loam	66	34	23	17	11	17
Heavily dunged arable loam	62	38	30	20	8	18
Pasture soil	53	47	40	22	7	25

THE SURFACE SOIL

The soil clods in the field are not very stable and can be broken down by rain or frost under certain conditions. They should not, however, fall down too easily into a fine mud, for in that case the surface of the soil usually dries into a hard crust unfavourable to the growth of young plants. The crumbling of these surface clods when they are wetted with rain can be controlled to some extent. Addition of farmyard manure or certain easily decomposable organic manures, or, if the soil is acid, of lime, often reduces the amount of fine mud produced when the clods are wetted, thereby either weakening the surface crust or preventing its formation. The improvement thus effected can be measured by sieving the dry soil in water. The method used at Rothamsted is to take a sample of soil from the field with the minimum disturbance of its structure, and allow it to become air dry. Then put between 50 and 100 gm. on a filter paper and slowly wet it by capillarity; after leaving for about 3 hours transfer it to the top of a bank of sieves[1] placed in water. The bank of sieves is then lifted out, and the water is allowed to drain away: the bank is carefully immersed again, and once more pulled out; again most of the water is drained away. This is repeated a number of times, making sure that no soil is pushed over the top of the sieves during immersion. The soil on the sieves is then dried and weighed.

The results show that farmyard manure, frost, or a previous ley usually improve considerably the stability of the structure, as is shown by an increase in the weight of soil on the sieves or by a reduction in the weight of soil passing through the finest sieve.

THE SUBSOIL

The lower portion of the soil differs so much from the surface layer that it receives a separate name, and is called the subsoil. The difference lies in the fact that neither plant nor animal life has been able to exert any great effect, so that the subsoil contains very little more than the original mineral material. It contains less organic matter than the surface soil, and in consequence has not the satisfactory physical properties conferred thereby and it possesses less nutrient plant material. The lack of organic matter is shown by its lighter colour.

Usually also it contains more clay and this, though it may be crumbly when dry, may become very sticky when wet. Both causes combine to render the subsoil less tractable than the surface soil, and on heavy soils it may become so bad that it is unwise to bring more than a little to the surface. Indeed, many acres of heavy

[1] Suitable sizes of mesh for a bank of four sieves are 3 mm., 1 mm., 0·5 mm. and 100 meshes to the inch. In working with stony or very sandy soils, the amount of stones or coarse sand on the sieves should be subtracted from the weight of soil on them.

land have been ruined by too deep steam ploughing which has buried the surface soil and left the plants with only a sticky unkindly subsoil. An advantage of the Norfolk method of ploughing "on the square", and of the one-way or turn-wrest plough, is that these do not, like ordinary ploughing, leave furrows of barren subsoil throughout the field between each of the lands. But they can be used only where the natural drainage is sufficiently good: on heavy soils the lands or "stetches" with their intervening furrows are commonly advantageous.

PLOUGH SOLES AND PANS

At a depth of a few inches below the surface of the soil there is sometimes a hard layer through which penetration is difficult. This is called a plough sole; it is the result of persistent ploughing to the same depth; the weight of the plough pressing on the bottom of the furrow causes a certain amount of consolidation which in course of time becomes very marked.

In acid soils where water is near the surface a hard layer known as a pan sometimes occurs. It is formed from dissolved material by precipitation of iron and aluminium oxides, silica and organic matter, in the form of a jelly which subsequently dries and hardens.

Plough soles and pans should be broken by occasional deep cultivation so as to allow free movement of the water and deep development of plant roots. This is best done by means of a subsoiler which breaks through the pan without burying the surface soil.

Once broken the pan must not be allowed to re-form, and for this reason the acidity of the soil must be neutralised by adding lime (p. 126).

CHAPTER III

THE ORGANIC MATTER OF THE SOIL AND THE CHANGES IT UNDERGOES

THE organic matter of the soil arises from the remains of previous generations of plants, animals and micro-organisms and can roughly be sorted out into two divisions according to age: some of it is as old as the soil, having been deposited along with the mineral particles when the soil was first made; some of it is newer, representing the residues of recently living plants. The older organic matter probably has little, if any, effect on the soil; its properties have been studied by examination of the subsoil at a depth below the range of the surface vegetation.

The more recent material is of supreme importance to the soil. It is subdivided into (*a*) the newest of all, the undecomposed roots or stubble which still retain some of the structure of the plant; and (*b*) the partly decomposed material, dark brown or black in colour, which has lost its structure and become completely intermingled with the soil; this part is commonly called "humus". The undecomposed part serves two purposes: it is the source from which the humus is derived, and it keeps the soil open and porous, maintaining passages through which water can drain away and air can enter, and preventing the mineral particles from settling down too compactly. In glasshouse practice it is always desirable to keep up the supply, and this is done by mixing good fibrous turf or, failing that, straw, in the borders (e.g. cucumber borders) so that the soil may remain open and aerated in spite of the constant heavy watering. After a time the fibrous material disappears and then the soil is much more likely to become sodden, covered with green growth, and "sour", than it was while the fibre lasted. In outdoor horticultural work it is equally an advantage to have sufficient undecomposed or fibrous material to keep the soil open, and afford what the gardener calls a proper root run. On heavy farm soils, also, undecomposed material, such as stubble, straw, long manure, is very helpful for the same reason. On the other hand, this material is a disadvantage on light soils because these are already open enough especially in dry seasons. Any fibrous or undecomposed plant material or manure containing long straw or peat moss is therefore added in autumn so that it may have a good chance of being broken up before the summer droughts come on.

The fibrous material contains many of the chemical substances that occur in the plant: among them are proteins, lignin, the celluloses and similar compounds, and waxes. The proteins break down to form ammonia and other substances, and the lignin and the celluloses give rise to the black mixture humus; carbon dioxide is given off during the process and other products are also formed. The waxes disappear only slowly; they tend to accumulate on soils like old garden soils to which much plant matter is added, and they are probably partly responsible for the curious difficulty in wetting these soils when once dry; drops of water tend to stand on the surface and not to soak in.

HUMUS

The mixture known as humus plays a specially important part in soil productiveness. In days gone by it was regarded as a distinct chemical group and was subdivided into humic acid, ulmic acid, etc., but these are now known not to be true chemical substances.

Humus as a whole cannot be separated from the soil like the sand or the clay. It can be destroyed by igniting the soil, when the black colour goes entirely and a reddish mass is left quite different from the original soil. The heat, however, has rather a drastic action and affects the mineral as well as the organic part of the soil. A gentler method of removing the humus is to oxidise it with hydrogen peroxide; about 50 to 100 c.c. are needed for 10 gm. of soil, and when the reaction slows down the mixture should be warmed. This time there is no reddening of the soil but the dark colour goes.

Part of the humus, however, can be extracted by means of dilute alkalis. Shake 100 gm. of soil with 500 c.c. of 5 per cent hydrochloric acid, allow to settle, pour off through a filter, and wash with water. Then transfer the soil to a bottle, add 500 c.c. of 5 per cent caustic soda solution, shake, and leave for some hours lying on its side so that as large a surface as possible is exposed to the alkali: shake periodically. Before long the alkali becomes dark coloured. Again allow to settle and siphon off, or, if you can, filter on a Buchner funnel by the aid of a pump: this is rather a slow process. To the clear dark-coloured filtrate add some strong hydrochloric acid drop by drop till the liquid is just acid.

A dark brown precipitate is thrown down containing part of the organic matter. On drying, this shrinks very much to little lumps almost black in colour which readily burn and leave behind a little red ash. Its composition varies considerably, but after it is thoroughly dried in a steam oven it usually contains about 50–57 per cent of carbon, 35 per cent of oxygen, and 3–8 per cent of nitrogen.

It does not appear that this "soluble humus" is of particular value in soil fertility though humus as a whole is highly important.

The physical effects of "humus" will be illustrated in Chapters VI and XIII; they fall under three headings:

(1) The organic matter imparts a black colour to the soil unless much chalk is present, when the white colour persists. It is a well-known physical law that a black substance absorbs more heat than a white one placed under similar conditions, hence the dark colour facilitates the warming of the soil in spring, when a rise in temperature of one or two degrees may be very important for the crop.

(2) The organic matter greatly increases the capacity of the soil for holding water. A soil rich in organic matter is, throughout the summer and autumn, distinctly moister than a soil poor in organic matter (p. 241).

(3) Organic matter facilitates the production of a fine tilth and a good seed bed, and it renders cultivation more easy. Soils well supplied with organic matter are therefore very valuable to the agriculturist both by reason of the large amount of nitrogenous substances they contain and also because of the ease with which they can be worked. Examples occur in the Fen

districts in this country, in the prairies of Western Canada, the black earths of Russia and elsewhere. Wherever they occur these black soils are readily taken up by farmers for cultivation.

There is, however, another type of organic matter which is less widely distributed and much less useful. Peat is organic matter but it is too acid and not sufficiently decomposed to be of much value, hence peaty soils are not in high agricultural repute. Intermediate between peat and fen comes another type found in the carr soils, which can be made distinctly useful by dressings of lime.

AMOUNTS OF ORGANIC MATTER AND OF NITROGEN IN SOILS

Owing to the great importance of the organic matter chemists have made many attempts to determine how much is present in the soil. Advantage is taken of the fact that organic matter burns away while mineral matter does not: hence some of the soil is burnt, and the loss of weight is measured. This method is simple, but unfortunately it is not quite sound, for the loss of weight includes some of the water that is very firmly held and also carbon dioxide and other substances given off by some of the mineral matter.

A better method of discovering how much organic matter there is in the soil is to determine, by a combustion method, the percentage of carbon present. Neither method, however, discriminates between the undecomposed and the decomposed material which, as we have seen, behave very differently in the soil.

The nitrogen content of the soil is an important factor; it gives an indication of the extent of the nitrogen reserves in the soil though it does not show how far they are useful to the plant. Table II gives typical examples; there is no very clear connection between the productiveness of the soil and the percentage of nitrogen or the loss of organic matter on ignition.

Table II. *Percentage of nitrogen and organic matter in typical soils and subsoils*

	Grass-land soils	Fertile arable soils		Poor arable soils		Barren wastes
Surface soils:						
Loss on ignition	8·6*	4·65	6·58	4·13	6·23	5·81
Nitrogen	0·27	0·12	0·22	0·13	0·14	0·17
Subsoil:						
Loss on ignition	—	3·00	4·94	3·74	5·50	2·70
Nitrogen	0·09	0·08	0·14	0·11	0·10	0·06

* Carbon (C) 3·26 per cent.

Nevertheless the results are of much value to the agricultural chemist in investigating soil fertility problems.

CHANGES IN THE ORGANIC MATTER

We must now turn to the changes undergone by the organic matter. During the process of cultivation the organic matter becomes oxidised and some of it disappears as gas; it thus suffers much more rapid changes than the mineral particles. Illustrations can be seen in the fens where the loss of organic matter has led to considerable shrinkage of the soil, and also in parts of North America where the original prairie soil was fairly rich in organic matter but after years of wheat cultivation it lost much of its stock. In Minnesota Snyder found that 50 per cent of the initial store of nitrogen was lost in twenty years' cultivation: in Saskatchewan Shutt observed a loss of 30 per cent of the nitrogen after a similar period. With the organic matter is lost also the advantages it conferred: the soil becomes impoverished and liable to erosion (p. 57): and, if much clay is present, it becomes difficult and expensive to cultivate. Hence such soils tend to go out of cultivation.

Losses also occur in market gardens and wherever large dressings of farmyard manure are applied.

A balance sheet drawn up for the soils on some of the Broadbalk wheat plots shows a far greater loss of nitrogen from the plot receiving farmyard manure annually than from those receiving artificial fertilizers only (Table III).

Table III. *Nitrogen balance sheet, lb. per acre, top 9 in. of soil, Broadbalk, Rothamsted*

	Farmyard manure Plot 2 B	No manure Plot 3	Complete artificials including 86 lb. N Plot 7
N in soil, 1865	4850	2960	3390
N in soil, 1914	5590	2570	3210
Difference in 49 years (+ = gain by soil, − = loss)	+740	−390	−180
Per annum	+ 15	−8	− 4
N added to soil *each year* in manure, seed and rain	208	7	93
N removed in crop	50	17	46
Difference	+158	− 10	+ 47
Deduct annual gain by soil	15	− 8	− 4
N unaccounted for each year	143	(Gain 2)	51

The carbon of the organic matter escapes as carbon

dioxide and some of the nitrogen is changed to ammonia; some is apparently lost as gaseous nitrogen.

NITRIFICATION

The ammonia remaining in the soil is at once seized upon by certain soil bacteria, the *Nitrosomonas* and others, and converted into a nitrite, and this is taken by another group of organisms, the *Nitrobacter*, and converted into a nitrate; the process is called nitrification. Thus the ammonia actually appears as nitrate which is readily detected in the soil. The amount of nitrate is commonly stated as so many parts of nitrogen per million parts of soil; they can be expressed as parts of nitrate of soda by multiplying by 6, or they can be converted into lb. per acre in the top 9 in. by multiplying by $2\frac{1}{2}$; the results are not quite accurate but suffice for purposes of comparison.[1]

The amounts of nitrate commonly found in arable soils at Rothamsted and Woburn are given in Table IV.

Table IV. *Amounts of nitrate found in arable soils at Rothamsted and Woburn*

	Expressed as nitrogen			Expressed as nitrate of soda
	Parts per million		Lb. per acre	
	0–9 in.	9–18 in.	0–18 in.	Lb. per acre 0–18 in.
Sand	5	4	25	150
Loam	10	8	46	276
Clay	10	6	38	228

In our climate nitrates do not accumulate to any great extent in the soil, and it is very unusual to find more than 24 parts per million or 120 lb. per acre (expressed as nitrogen) in the top 18 in. It sometimes happens in

[1] More precise figures are: 1 part nitrogen = 6·06 parts nitrate of soda.

dry regions that higher amounts are present, but it is usually supposed that they got there by evaporation of water which had soaked in from somewhere else, concentrating the nitrates from a wide area over a particular spot.

Under our climatic conditions the nitrates have no opportunity of persisting long but are either washed out by rain or taken up by plants. Once the stock is reduced a further quantity begins to be formed, and so far no limit has been reached to the amount of nitrate a soil can be made to yield. One of the Rothamsted plots which has been cropped with wheat every year since 1843 and has had no manure since 1839 still goes on yielding nitrate, and Table V shows the amounts of nitrate nitrogen contained in its various sections in November 1935.

Table V. *Nitrate nitrogen in soils taken in November* 1935 *from the unmanured plot, Broadbalk*

	Nitrogen, lb. per acre			Equivalent to pure nitrate of soda, lb. per acre		
Depth	Section I	Section II	Section III	Section I	Section II	Section III
0–18 in.	30	23	35	180	138	210
0–27 in.	42	33	53	252	198	318

Section III had been fallowed during the summer: the others had been cropped.

Another piece of land is kept bare of all vegetation and is undermined in such a way that the whole of the drainage water can be collected for analysis. Ever since 1870 when the experiment began the land has yielded a large supply of nitrate, the amount being equivalent to 300 lb. of nitrate of soda per acre every year for the first 29 or 30 years, and to some 170 lb. in more recent years.

DENITRIFICATION

A further change goes on in certain circumstances. When all air is excluded from the soil by flooding it for a long time with water, the nitrates are liable to decompose, yielding nitrites and subsequently gaseous nitrogen. This change, known as denitrification, only goes on slowly in cold weather and probably is of rare occurrence under British agricultural conditions where land would only be waterlogged in winter, if at all. But it seems to go on in the wet rice fields of the East and in these circumstances nitrates are not used as manure, but ammonium salts or oil cakes (p. 212).

NITROGEN FIXATION

All these changes result in loss of nitrogen: fortunately there are others that bring about gains, chief among them being the fixation of gaseous nitrogen by the organisms in the root nodules of leguminous plants. This process proceeds vigorously during the growth of clover, trifolium, lucerne, sainfoin, vetches, etc., and these crops therefore enrich the soil considerably.

Some nitrogen is also fixed by certain free-living bacteria called *Azotobacter*. These require considerable quantities of decaying plant residues as a source of energy; for the process is not one that will continue by itself once it is started, like the burning of a bonfire, but rather it resembles hauling a load up a hill and requires the continuous application of energy. Both processes take place in land laid down to grass. The gain does not go on indefinitely: after a time it is counterbalanced by losses; but the net result is that grass land contains considerably more nitrogen than arable land (p. 46). The nitrogen comes *from the atmosphere*, and thus represents an absolute gain to the soil. The following simple rule should be remembered: land in sod gains *nitrogen*, land in fallow gains *nitrate*. The gain in nitrogen is absolute, but the gain in nitrate is not, it only represents a change of one form of soil nitrogen into another. When grass land is ploughed up the gain in nitrogen ceases, and the gain in nitrate begins; the stored up nitrogen can thus come into service for a series of arable crops. This is the advantage of long leys and alternate husbandry: during the ley period fertility is stored up; during the arable period it is cashed. Sufficient may be produced to yield corn crops so heavy as to justify the ploughing up of pasture which is not of first-rate grazing quality.

Students sometimes wrongly confuse nitrification with nitrogen fixation, and use the word "dentrification" to mean all losses of nitrogen.

THE SOIL POPULATION

This whole chain of processes is of great importance to soil fertility because it brings about the conversion of the undecomposed plant residues, which are of little value to the soil except to open it up, into valuable humus material and plant food. The process is a release of stored-up fertility, and it has to be encouraged by every means in the cultivator's power. It is mainly brought about by living agencies; earthworms play a preliminary part by dragging the materials into the soil and effecting a proper admixture; but moulds and bacteria are the important decomposing agents. Fortunately all these organisms require substantially the same soil conditions as plants: air, water, proper temperature, and food, absence of injurious substances, etc. The soil population is, however, very complex and the organisms are not all useful to plant growth; some indeed are positively harmful: land left long in grass accumulates wireworms which do a great deal of harm to the arable crops grown after it is ploughed out. Land frequently carrying one and the same crop sooner or later accumulates any soil, animal, insect or fungus pest to which the crop is liable: e.g. eelworms on potatoes and in many countries on sugar beet, foot-rots (e.g. "Take all") on wheat. Methods are being worked out for controlling the soil population so as to encourage the useful forms and repress the others. Partial sterilisation by heat, mild antiseptics, drying in the sun, etc., increases the productiveness of the soil. These treatments also kill insects and some at least of the disease organisms. Practicable methods of applying them have been worked out for horticulture but not yet for agriculture.

SUMMARY OF CHANGES IN THE ORGANIC MATTER

We may now summarise the processes described in this chapter.

Decomposition of the dead plant residues in the soil plays a highly important part in soil fertility:

(1) The cellulose and lignin give rise to humus.

(2) The proteins give rise to ammonia, which is then oxidised to nitrates. The production of ammonia is called "ammonification", and the oxidation to nitrates "nitrification".

(3) The nitrates are taken up by growing plants and built up into protein. Certain soil organisms can effect the same change in presence of carbohydrates.

(4) In absence of air some of the nitrates are reduced

to gaseous nitrogen which escapes. This process is called denitrification.

(5) A loss of nitrogen also occurs when rich soils are cultivated. It is possible that this is also denitrification but the evidence is not yet sufficiently definite to justify the use of the term.

(6) The losses indicated in the preceding paragraphs are made good in natural conditions in two ways: (*a*) organisms associated with leguminous plants are able to fix gaseous nitrogen from the air; (*b*) in presence of easily oxidisable organic matter, especially of carbohydrates, certain organisms are able to fix gaseous nitrogen and build it up to the form of protein. This process is called nitrogen fixation; it must not be confused with nitrification.

The protein formed during nitrogen fixation, however, can undergo nitrification in the usual way, being decomposed to form ammonia, which is then oxidised to nitrate.

These processes form a double cycle in the soil which may be thus expressed:

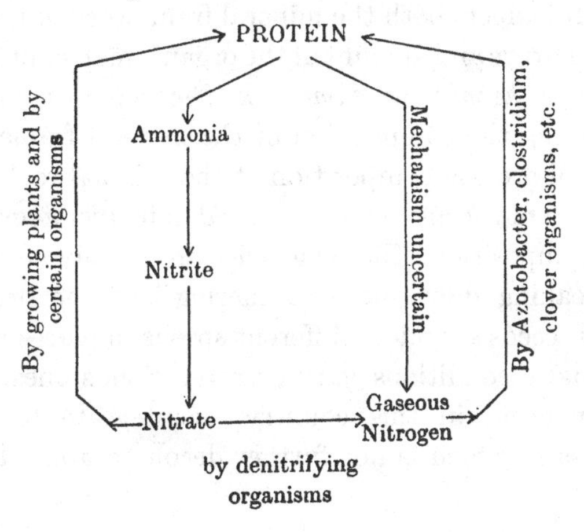

CHAPTER IV

THE EFFECT OF CLIMATE ON THE SOIL AND ON FERTILITY

THE horticulturist working under glass can have any soil and almost any climate he likes to pay for, but the farmer must accept the climate as he finds it and put up with any effect it may have on the soil and on the crop; the best he can do is to ascertain what these effects are and then prepare against them. They fall into three groups:

(1) Climate helps to make the soil and decide its type.

(2) It influences the productiveness of the soil.

(3) It determines the crops that can be grown, and hence is the final agent to decide what shall be done with the soil.

THE EFFECT OF CLIMATE IN DETERMINING THE GENERAL CHARACTER OF THE SOIL

Climate affects both the mineral framework of the soil and the nature and amount of the organic matter present.

Effect on the mineral framework. The two great factors determining the composition of the mineral framework of the soil are the composition of the original rock, and the nature of the agencies concerned in its disintegration and decomposition. These agencies are mainly climatic. The breaking down or "weathering" of the original rocks proceeds at widely different speeds in places where the climatic conditions vary greatly. Sandstones, such as often occur in this country, disintegrate to form quartz sand which is not further decomposable. Limestone dissolves away altogether and may leave great gorges or caves. But the more complex rocks go through a series of decompositions which take place most rapidly in regions of high temperature, high rainfall and good drainage.

Climate regulates the composition of the soil in a second direction. As we have already seen, the particles formed from the rocks do not remain where they are but get carried away by various climatic agencies such as running water, ice, or wind. Usually there has been some selection and the particles have become sorted out to some extent and suffered changes on the journey. The amount of sorting and the extent of the change depend largelv on climatic factors.

Effect on the organic matter. The mass of mineral particles formed by weathering of the rocks and sorting out by subsequent agencies is only the framework of the soil. It soon covers itself with vegetation which gradually converts the mineral mass into a true soil.

The character of the soil is very much affected by the nature and amount of the organic matter present, and this is largely determined by the type of vegetation that grows there and the extent to which the decomposition has proceeded; both these are climatic effects. Under dry conditions the plants tend to be narrow-leaved and tough—e.g. pine needles, broom, etc.—whilst under moister conditions a larger more leafy type of vegetation arises. These two types of vegetation break down in very different manner in the soil: the large leafy plants yield a large supply of useful humus material, while the shrubbier and more leathery plants of the dry situation do not. There may be plenty of organic matter in these dry soils; the light dry sands of the Sussex heaths sometimes contain as much as 10 per cent, but it exists in the form of undecomposed bracken fronds and similar residues, and is of no agricultural value because it is not properly decomposed.

SOIL LOSSES

So far we have been considering only the building up of the soil; we must now study the losses that are going on. The processes that called the soil into being are still operative to-day; the transport of the material did not come to an end when the soil was brought into its present position but continues, and tends to remove the soil now that it is formed. The losses have gone on simultaneously with the formation of the soil and they still continue. In temperate climates such as Great Britain the most important source of loss is the rain. As rain falls on to the land and soaks in, it dissolves out soluble substances and carries them away. Hence the drainage waters are often hard and may be unfit for drinking. The constituent removed in largest quantity is calcium carbonate. The amount of loss depends on the quantity present, but on uncropped soil it not uncommonly varies from 100 to 300 lb. oxide of lime (CaO) per acre per annum. Other constituents, especially nitrates, are removed in smaller but still important amounts. Thus in course of time a soil exposed to a heavy rainfall tends to become impoverished through loss of plant food, and acid or "sour" through absence of calcium carbonate. These losses, however, are diminished when the soil is covered with vegetation: they are at a minimum in grassland.

SOIL EROSION

In the dry regions of the world another kind of loss frequently occurs. The natural vegetation is usually slow growing, and once it is removed, either by cutting down the forest or by ploughing or overgrazing the steppe or prairie, it does not easily recover. The soil is thus left completely exposed to wind and rain. In dry regions the winds are frequently high and the rain when it comes is apt to be torrential, hence both wind and rain carry away the soil: the process is called soil erosion. Less than a hundred years ago there were vast areas of fertile prairie land in those parts of western North America where the average rainfall varies from about 22 to 15 in. annually. The first settlers soon found them to be very suitable for wheat growing and they developed a system in which wheat alternated with fallow, this being well cultivated to conserve the soil moisture (Fig. 11). Nothing was done to return organic matter to the soil, and constant cultivation so reduced its amount that the soil gradually became more and more powdery till finally it was so fine that the wind could blow it away. Great dust storms frequently arose which removed the soil from millions of acres, leaving only a desolate waste where once there had been farms (Fig. 12).

Rain brings about soil erosion in two ways: sheet erosion, in which thin layers of soil are washed away from large areas; and gully erosion, where the soil is scooped out to a depth of several feet from a more limited area (Fig. 13). Notable examples of gully erosion are found in South Africa, brought about, not by

Fig. 11. Harrowing in Western Australia as part of wheat cultivation. The rainfall being low, water conservation is very important.

Fig. 12. Dust storm in Western Canada. The continued cultivation has reduced the soil to so fine a state that it is readily blown away by the wind leaving the region barren. This is called wind erosion.

constant cropping, but by overgrazing. In the old days when for a variety of reasons the cattle population was low the grazing problem was not serious. But in the last 50 years the cattle population has grown considerably. The vegetation cover was not very dense and it was grazed

Fig. 13. Washing away of soil by torrential rainfall, United Provinces, India. This is an example of water erosion.

hard. The climate was, on the whole, dry, with a long dry season, followed by rain which might be torrential. Usually on the pastures the water supply for the animals was restricted to a few places. The animals frequently, in accordance with their custom, walked in single file, so compacting the soil and making long shallow depressions. When torrential rain came, it started running off the land along these depressions, carrying the soil with it, gathering force as it went, and finally scooping out great gullies called dongas. Matters were made worse by the burning of the grass, which was done periodically to destroy animal parasites, especially ticks, and also to ensure a growth of good young grass for the animals.

Overgrazing is one of the most serious causes of erosion in Australia.

A third type of soil destruction took place in certain tropical countries, e.g. in Ceylon. Here there is no dry season, but a high rainfall which is sometimes torrential. Vegetation grows well and a cover is quickly established. In the hill regions, tea and coffee plantations were started; the trees and bushes were cut down, the tea or coffee plants were planted, and, according to good western European practice, the plantations were kept nicely hoed and free from weeds. Then erosion began: the heavy rain washed the soil away and serious destruction set in. Clean weeding is now recognised as bad, and selective weeding is being tried, or else the planting of some crop that will give cover, but not compete too seriously for soil nitrogen. In Malaya also, this "clean culture" is being given up in favour of a "forestry system", where much undergrowth is left.

Deforestation is a very potent cause of soil erosion, and it has operated in many places, because the nineteenth-century migrants seem to have been prejudiced against trees and regarded them as pests to be destroyed rather than as allies to be preserved.

Another source of damage almost as harmful, particularly in India, Cyprus, Palestine and Africa, is the grazing of the forest by domestic animals. So long as animal diseases and pests, tsetse fly, etc., and wild animals abounded, the animal population did not usually become numerous enough to do much damage, and tribes were often given rights of grazing which were at the time unobjectionable. But with the development of a peaceful life and an efficient veterinary service, the animal population has increased so much that the forests are badly damaged, and the growth of human population necessitates the clearing of more forest for the planting of food crops.

THE IMPORTANCE OF VEGETATION FOR SOIL CONSERVATION

Erosion is serious only when the vegetation cover is removed, and although this sometimes happens through natural causes it is much more frequently brought about by human activity. The soil and its vegetation cover should be regarded as inseparable for soil conservation purposes. We have already seen that the plants growing on the soil supply it with an essential constituent, organic matter. But they do much more. They absorb and transpire considerable quantities of water, thus reducing the loss of soluble material which would have taken place had the water soaked through the soil. They absorb plant food which might otherwise be lost, storing it in their roots, stems, and leaves, and returning it to the soil when they die unless in the meantime they have been taken away by men or animals. While they are growing, their roots exert some action on the soil, causing the fine particles to unite into crumbs which constitute the best structure for soil fertility and soil conservation. And, finally, growing plants directly protect the soil against rain and wind and so reduce the amount of loss that might occur. The role of the vegetable kingdom is essentially one of construction and conservation.

REMEDIES FOR SOIL EROSION

The remedies for soil erosion are often difficult in practice, though they are simple enough in principle. First, the movement of the soil must be stopped, the gullies and dunes must, if possible, be levelled, and a vegetation cover must be established—all frequently very difficult tasks.

The stopping of soil drift due to wind is done in two ways: by setting up wind breaks, and by planting some vegetation that will fix the sand or soil and stop its movement. A gigantic scheme for planting shelter belts on about a million acres of farm land within a zone 100 miles wide and stretching from Canada into Texas, has been begun in the United States. The planting of blowing sands with various grasses is now a well-recognised device with various ingenious modifications; e.g. Whitfield suggests the planting of only the lower parts of the dunes, leaving the upper parts loose and exposed, so that they might be blown into the hollows, thus making the wind do the levelling.

The stopping of water erosion is done by setting up obstacles to the running away of the water. The old method of doing this was by a system of terracing such as the very early or prehistoric lynchets on the chalk downs of England, or the terraces built up with stone walls in the Mediterranean countries, and in parts of India and Africa. On steeply sloping land this is very effective, but where the slope is less, an easier method can be used—ploughing and draining along the contours instead of up and down, also setting up of shallow banks to catch any soil that is being washed away. Another method, much used in India, is to set up bunds or banks across the depression along which the erosion will go. In other cases a simpler device is sufficient: it is necessary only to maintain a good vegetation cover and at regular intervals to plough two furrows along a contour. Whatever method is adopted the run-off of the water must be reduced as much as possible and steps taken to ensure that it soaks in rather than scours away the surface soil.[1]

RE-ESTABLISHMENT OF THE VEGETATION COVER

The most important step in dealing with soil erosion, however, is the re-establishment of the vegetation cover. The particular method adopted depends on the circumstances: it may be afforestation, sowing with grasses, or arable crops. The cropping system should give at least as good protection as the natural cover. This is very difficult where a single crop dominates the agriculture, because, with the exception of grass and forest, no single crop confers the necessary stability, and continuous wheat-growing is likely sooner or later to kill itself by sheer destruction of the soil. The best hope is for mixed farming, where both livestock and grain are produced, and where grasses and clovers necessarily play an important part. A good rotation on the arable land, fencing improvement of the grazing lands, rota-

[1] For fuller details of Soil Erosion see *Tech. Comm. Imp. Bur. Soil Sci.* no. 36, 1939.

tional grazing, the avoidance of overgrazing, and good manuring and management, afford the best preventives against erosion.

Work on soil conservation is proceeding actively in many parts of the British Empire. Ceylon, one of the first countries in the world to recognise the dangers of soil erosion, offers good examples of control measures suitable for European-managed plantations in the tropics. Southern Rhodesia affords examples of control in native areas: the problem is being attacked simultaneously from the mechanical side by ridging, contour cultivation, etc.; from the agricultural side by better manuring and improved rotations; and from the economic side by attempting to modify conditions of land tenure in a way that discourages exhaustive farming.

Soil erosion goes on in Great Britain though it does not usually trouble the farmer. The Roman city of Verulam at St Albans, once flourishing, is now buried under several feet of silt washed down from the higher ground, so that it is necessary to dig down to reach the old surface.

SOIL BELTS AND CLIMATIC ZONES

We have seen that right from the commencement the soil has been moulded by the climate, and it is not surprising, therefore, that in parts of the earth with characteristic climates the soils should also have certain definite features. In classifying soils it is necessary first to divide them into great groups according to the climatic conditions under which they were formed and then to subdivide these groups according to the geological origin of the material and other properties.

These zones can most easily be recognised in any great continental area. In the great dry belt in the middle west of North America there is a scarcity of vegetation, consequently no large amount of organic matter finds its way into the soil. Further, soluble salts frequently occur in the soil; some are directly harmful to vegetation while others indirectly injure it by depriving it of such little soil moisture as is present—for plants can only take water from weak and not from strong solutions. The salts arise in part from the breaking up of certain mineral particles, but in the main from pre-existing inland seas or lagoons that have long since disappeared. Soils thus charged with salts are called alkali soils; they occur sometimes in patches and sometimes in great areas, but are always dreaded alike by cultivators and travellers. For they yield no drinking water, and as they dry the wind blows them up into the eyes and mouth and nostrils causing the membranes to smart. In the absence of rain the soils must be irrigated, and while the irrigation water washes out the salts from some areas it carries them on to others, killing the vegetation. The soil is curiously mottled in appearance: it forms hard white lumps around which black water collects or dries to leave a black crust behind. It is hard on top but often mushy below, especially in irrigated regions, and below the surface is a thick stodgy clayey mass. Proper drainage and in certain circumstances treatment with gypsum have done much to reclaim these lands.

Moving eastwards and northwards there is a rather moister belt with more grass and less salt, but the vegetation is still wiry or leathery and there is no great amount of organic matter in the soil. These are the steppe soils which can be found in parts of the western States and of Alberta. Alkali spots still occur, and Fig. 14 shows one on a farm at Fremont, Nebraska.

Farther eastwards and northwards is a zone of higher rainfall where the conditions were such that more organic matter accumulated. The framework is often a

Fig. 14. Alkali spot, Fremont, Nebraska.

loess, a wind-carried soil derived from glacial drift, frequently mingled with calcareous delves. For long periods of years the grasses and other plants (not usually trees, the rainfall being too low for forest) grew almost undisturbed, and their roots caused a marked accumulation of organic matter in the soil. In places there are great areas of old lake bottoms, the soil of which contains even more organic matter. These soils form the great wheat belt of western Canada and the United States, and the maize belt of the south. The winters are clear and cold and the summers hot and dry. The total rainfall may be sufficient, but its distribution is not always favourable to maximum crop production (see Fig. 21, p. 85).

Still farther east are regions of wood and forest where the climatic conditions approximate more closely to our own; the soils also resemble ours in England.

Fig. 15. Damage by salt in an orange plantation in the Murray river irrigated region, Australia. The trees have failed in the patch that is affected though they are still growing well around it.

In wetter, cooler districts much of the rain water soaks through the soil, washing out calcium and other bases and finally leaving the soil acid, which causes considerable change in its composition. Organic matter tends to accumulate at the surface; as the acid water soaks through the underlying soil it dissolves out calcium, aluminium, iron and other oxides, leaving finally only a bleached layer of inert material very poor in plant food. Some of the clay disintegrates and decomposes, and is also washed down. Some of these substances, including the nitrate, some of the calcium carbonate, and a little of the iron and silica, are carried away to the rivers and so to the sea. The other substances are not carried far but are deposited in the subsoil forming an enriched layer of darker colour. The process goes farthest under coniferous forest or moorland plants: the soils ultimately formed are called podsols, a Russian name, because they occur frequently in north Russia and have there been much studied. Examples can be found in Great Britain, but here partial podsolisation is more frequent.

A wholly different type of soil, known as the tundra, is found in the far north in the barren lands. It is black and peat-like and the subsoil is as a rule permanently frozen: it is covered only with mosses, lichens, etc., and lies beyond the regions of our accustomed vegetation. These various soil types are shown diagrammatically in Fig. 16, drawn up by E. M. Crowther.

Any other continental area can similarly be divided into zones corresponding broadly with climatic zones. In Russia, for example, white desert soils poor in organic matter but often containing alkali are to be found in the dry region north of the Caucasus: farther north under a limited rainfall of 8–12 in. occur the brown steppe soils, their deeper colour indicating their higher content of organic matter; pushing still farther north a belt of chestnut-coloured soils is found stretching away in a north-easterly direction from Podolia in the south-west across Little Russia to Samara and Orenburg in the east. Above this again comes the famous belt of black earth, the Chernozem, the nearest European approach to the black soils of the western prairies and like them devoted largely to the cultivation of wheat; these are found in Hungary and continue north-easterly

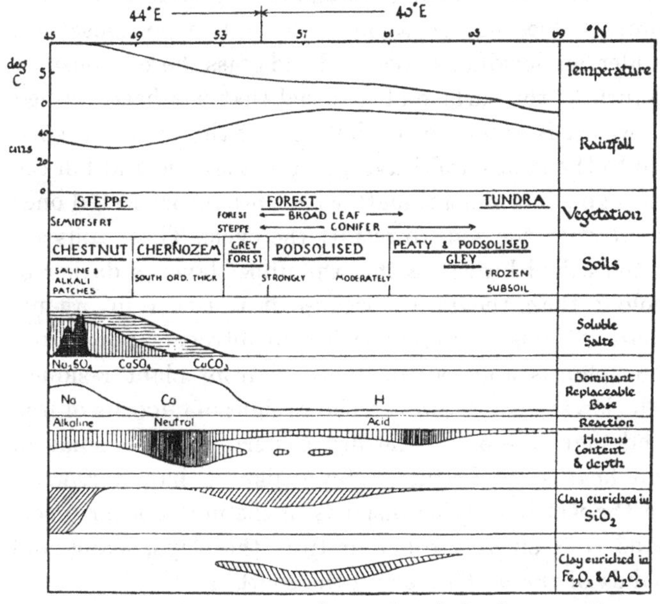

Fig. 16. Showing for a continental area the relation between climate, vegetation and soil type. The lower half of the diagram gives some of the characteristic features of the various soil groups (E. M. Crowther).

through the west Russian province Volhynia to the Government of Perm. Farther north these are succeeded by the podsolised forest soils, and these in turn by the podsols, white, poor, acid soils in a cold wet belt of coniferous forest; and finally above them come the tundra soils, acid, treeless, carrying only lichens and mosses.

CLIMATE AND SOIL IN ENGLAND

Even in England indications of climatic zones can be traced, although in the main our soils fall into one great group of woodland origin. In our moist cool climate there is a good deal of washing downwards, but the process does not go as far as complete podsolisation; under our deciduous woodland and grass there is constant return to the surface of material that has been washed down: it is absorbed by the roots of the plants, carried up to the stems and leaves, and as they die and fall the nitrogen and mineral matter becomes incorporated once more with the soil, to restart the cycle. These soils are often called brown earths: the upper layer is darker in colour than the lower because it is richer in organic matter: it is, therefore richer in nitrogen, potassium, phosphorus and calcium derived from plant residues. The lower layer, however, shows signs of deposits of iron and manganese. In the dry eastern counties some of the heaths are distinctly steppe-like in character, while in the wet high-lying districts of the north occur moorland soils entirely different from the clays, loams and sands of the midlands and the south.

THE EFFECT OF WEATHER ON THE SOIL

While climate plays a great part in determining the general character of the soil, the weather is responsible for tolerably wide variations exhibited from year to year.

There are at least five ways in which the weather or seasonal effects operate:

(1) High rainfall tends to wash out two very useful

constituents, calcium and nitrates, both of which must be replaced or the soil loses fertility. Fortunately other useful substances are absorbed by the soil and are therefore less liable to be lost.

(2) High rainfall has an adverse physical effect and spoils the tilth.

(3) In dry conditions there is less or no washing out of calcium carbonate or of nitrates, and hence less wastage of fertility.

(4) Drought, frost, hot sunshine, and other factors which are detrimental to plant life are finally beneficial to bacterial activity (p. 52), and lead to an increased production of plant food.

(5) Frost and also drought and rewetting, have a beneficial effect on tilth.

These factors of course all intermingle in their action.

The washing out of calcium and other bases explains why the land at the top of a slope in a wet region is liable to be acid, while the land lower down is neutral and enriched from the soil above. Marked changes in soil reaction and in vegetation are observed in passing from the top to the bottom of the northern chalk downs in Bedfordshire and Hertfordshire. The bottom soil is deeper than at the top because there has been washing down.

The nitrates formed during spring and summer by bacterial action, and destined to serve as food for the next generation of plants, are readily washed out during a wet winter, but they remain safely locked up in the soil throughout a period of frost and snow when no leaching takes place. There they lie ready for use when spring awakens the young plants into activity; consequently a mild spring following on a hard winter is commonly a period of vigorous growth. This is well seen in Canada, where a remarkable development of vegetation takes place directly the weather is sufficiently warm. In part the result is due to the effectual cold storage of the plant food, neither loss nor deterioration going on in frozen ground, in part also to the increased activity already mentioned of the food-making bacteria after a spell of adverse conditions.

Another effect of a wholly different nature is also produced. Frost puffs up or lightens the soil: it splits the hard clods and brings them down to a nice crumbly tilth well adapted for a seed bed. Further, it tends to change clay from the sticky into the crumbly state. On the other hand, long-continued wetness has the opposite effect: it consolidates the soil, makes it sticky and very unsuitable for seeds. Thus at the end of a mild wet winter the soil is poor in plant food because of the leaching that has gone on, and it is in a bad mechanical condition because the wetness has made the clay particles very sticky. On the other hand, at the end of a more severe winter when the land lay frostbound or covered with snow there is a good supply of plant food, all the autumn reserves having been safely locked up in the soil, the conditions have been favourable for the rapid production of plant food, and the texture of the soil is very favourable for the production of a good seed bed. The advantages, therefore, are wholly in favour of a dry cold winter, and we can see the wisdom of the old proverbs:

> Under water famine, under snow bread.
> A snow year is a rich year.

Sir W. N. Shaw calculated that every inch of rain falling during the autumn months—September, October, and November—lowers the yield of wheat during the next season in the eastern counties by a little over 2 bushels per acre (2·2 to be precise) from an ideal standard of 46 bushels per acre.

The older writers, noticing the value of frost and snow, thought they had an actual fertilising value, and indeed many gardeners and farmers still contend that snow is a manure. Opinions of good cultivators are always entitled to respectful consideration, and many analyses of snow have been made, but they have failed to reveal any appreciable amount of fertilising constituents. Snow differs a little from frost in its action: it forms a non-conducting coat for the soil and prevents the temperature from falling as low as it otherwise would. All plants benefit by the snow cover because their roots are protected from excessive cold, while young plants such as winter corn gain by being shielded from drying winds.

A hot dry summer has at least as beneficial an effect on the soil as a cold dry winter. The drying out certainly changes a heavy soil into clods, but when these are moistened again by autumn rains they readily fall to a good tilth. Really hot sunshine, however, appears to have some bad effect on soils rich in organic matter, and in the fen districts a summer fallow is regarded as harmful, as also in certain tropical countries, e.g. Ceylon and Malaya.

THE EFFECT OF SEASON ON THE NITRATE CONTENT OF THE SOIL

The production of nitrates in the soil (which, as we have seen, is an indispensable process for the welfare of the crop) takes place most rapidly in our climate in late spring or early summer. The accumulation of nitrate slackens down while the plant is growing and rapidly absorbing nitrate, but it may speed up again in a warm

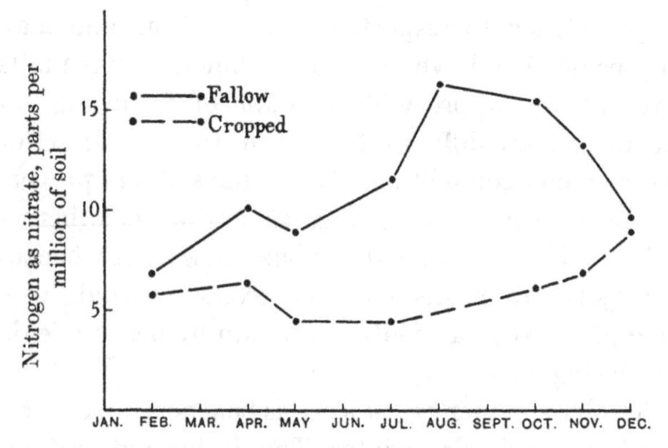

Fig. 17. Curve showing average amounts of nitrate present in cropped and fallow soils at different seasons of the year. (Rothamsted.)

moist autumn. Typical results are shown in the curve of Fig. 17. In a dry summer the nitrate formed is all left in the soil or taken by the crop: in a wet summer some of it is washed out. This is shown by comparing the amounts of nitrate present on an unmanured fallow plot at Rothamsted during the wet summer and autumn of 1912 with those present in the dry summer of 1913. In the top 18 in. of soil amounts were found equivalent to the following quantities of nitrate of soda, in lb. per acre, showing a very great difference in favour of a dry summer:

	Feb.	May	Sept.
Dry summer, 1913	126	312	376
Wet summer, 1912	180	138	114
Difference in favour of dry summer reckoned as nitrate of soda (lb. per acre)		174	262

The nitrates left in the soil at the end of September represent the initial stock for the farmer during the coming season. After a dry summer it is high, after a wet one low. How much of it ever gets into the crop depends on the winter weather. A wet winter will wash much of it out while a dry winter conserves it safely. During the wet winter of 1911–12 the following losses took place from some uncropped soils at Rothamsted, Ridgmont and Milbrook:

	Loam in good heart Rothamsted	Poor loam Rothamsted	Clay Ridgmont	Sand Milbrook
Present in Sept. 1911	690	306	234	102
Remaining in Feb. 1912	186	168	180	54
Lost during winter	504	138	54	48

Reckoned as lb. of nitrate of soda in top 18 in. per acre.

The loss from sand is small because the stock happens to be low, and from clay it is also small because percolation of water does not readily take place. The most serious losses are from good loams. In dry winters the loss is less, but not infrequently the loam at Rothamsted loses during winter months as much nitrate as would be required by a four-quarter wheat crop.

LOSS BY DRAINAGE

If the student has access to drainage-water from a field he should make the following experiments periodically:

(1) Test for nitrate and compare with a standard solution to ascertain approximately the concentration.

(2) Test for calcium. In many cases so much calcium bicarbonate is present that a precipitate is thrown down on warming the solution.

The following experiment shows how a crop affects the drainage:

Take two glazed tubulated pots (Doulton's "mixing jars" shown in Fig. 1), fill with soil, keep one pot fallow, sow grass-seed in the other. Fit the tubulure with a tight cork through which passes a glass tube bent so as to deliver the drainage water into a bottle. Measure the amount of drainage after rain and estimate the nitrate present. The experiment must run over the whole season; in a period of drought rain water may be *gently* supplied from a water-can, although it is hardly possible to simulate the action of rain itself.

The various bad effects of wet weather are reflected in the crop. A wet winter is notoriously bad for the wheat crop; on the other hand a dry winter is much more favourable. Shelter of course is just as effective as dryness: the ground where a stack has stood during winter is well known to be more productive than adjoining ground that has been exposed to the rain.

The practical point arises: how can the cultivator remedy matters? He must try both prevention and cure. Loss of nitrate can be prevented by sowing catch crops in autumn to be ploughed in or folded before the spring sowing (p. 136). Bad tilth can be diminished by leaving the ploughed land rough and taking care that the wheat land does not get too fine.

Loss of nitrate can be made good by spring dressings of quick-acting nitrogenous manures: nitrochalk or sulphate of ammonia if the surface is sticky, or nitrate of soda if the soil can be got into reasonable condition (Chapter VIII).

THE EFFECT OF CLIMATE IN DETERMINING WHAT CROPS CAN BE GROWN

The fertility of a soil is judged by its power of producing crops, but it obviously cannot grow crops unless the climate allows: we therefore have to turn to the effect of climate in deciding what crops can and what cannot be grown. There is a fairly simple connection between the type of crop and the climate. In general seed does not ripen well in wet seasons or districts, and crops wanted for the sake of the seed are usually grown in dry rather than wet districts. On the other hand, actual plant growth, i.e. growth of leaves, stems, and roots, is much better in moist than in dry districts or seasons. For example, the abnormally dry summers of 1934, 1935, and 1938 were excellent for grain crops so that the corn was uncommonly good, but so bad for the growth of grass that hay became scarce and dear. The wet summer of 1936 was very favourable to the growth of grass, swedes, etc., but bad for the production of seed. Another factor also comes into play. Very wet land cannot easily be dug or cultivated: it is therefore left in grass, which is much grown in wet districts. Since grass has to be used by animals of some sort a good deal of livestock is usually kept either for the production of meat, butter, cheese, etc., or for breeding young animals to be sold to other districts. Wherever cultivation becomes expensive for any reason there is a tendency to resort to grass and pastoral conditions. This system of husbandry can be much intensified by using good seeds mixtures, suitable manures, proper cultivation of the grass, and well-planned rotational grazing.

The following rules will be found useful in discussing crop production in temperate regions; they are, however, by no means absolute. Warm districts yield early crops, and are therefore well adapted for market-garden produce and for fruit. Moderately dry regions are suited for seed crops. Moister regions are adapted for seed crops that need not fully ripen, such as oats, for root crops like mangolds, swedes or potatoes, or for leaf crops like the cabbage tribe. Wet regions are commonly devoted to grass.

Fig. 18 shows the distribution of rainfall in England and Wales; Fig. 19 shows the distribution of wheat, and Fig. 20 that of grass. It will be observed that the wheat growing tends to concentrate in the drier eastern parts of England and grass growing in the moister west and north.

These same factors that determine the regional distribution of crops operate everywhere, and many illustrations of their action may be found within very restricted areas.

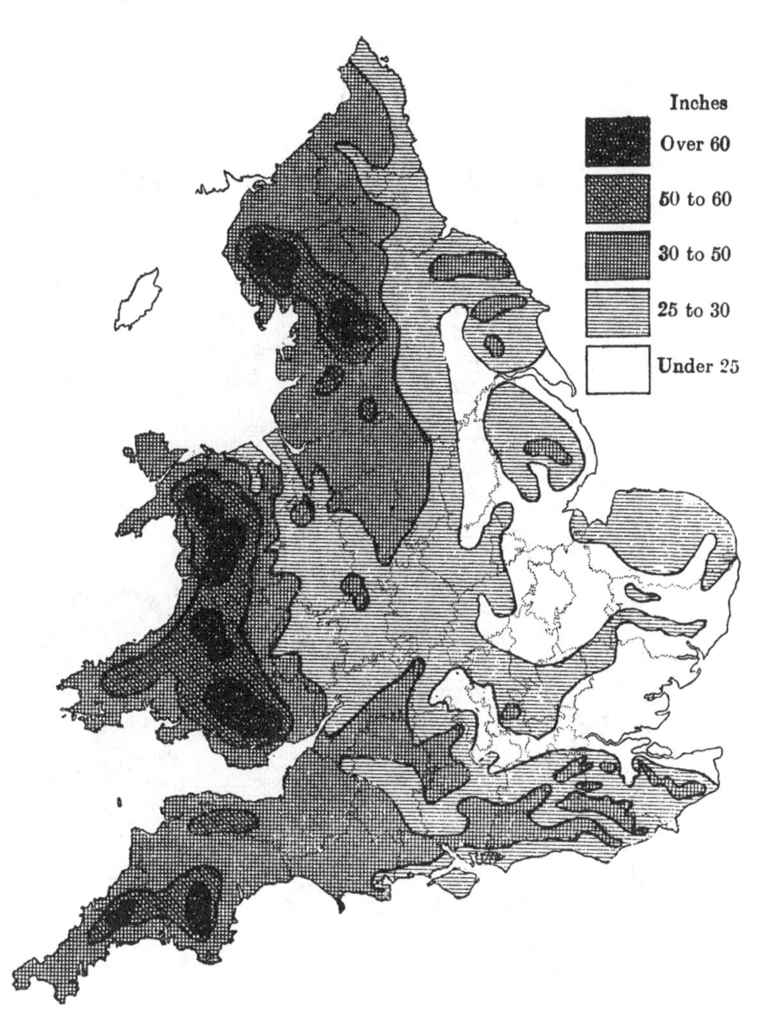

Fig. 18. Annual rainfall in England and Wales. For a detailed discussion see H. R. Mill and C. Salter, *Journ. Royal Meteorolog. Soc.* 1915, vol. LXI, p. 1.

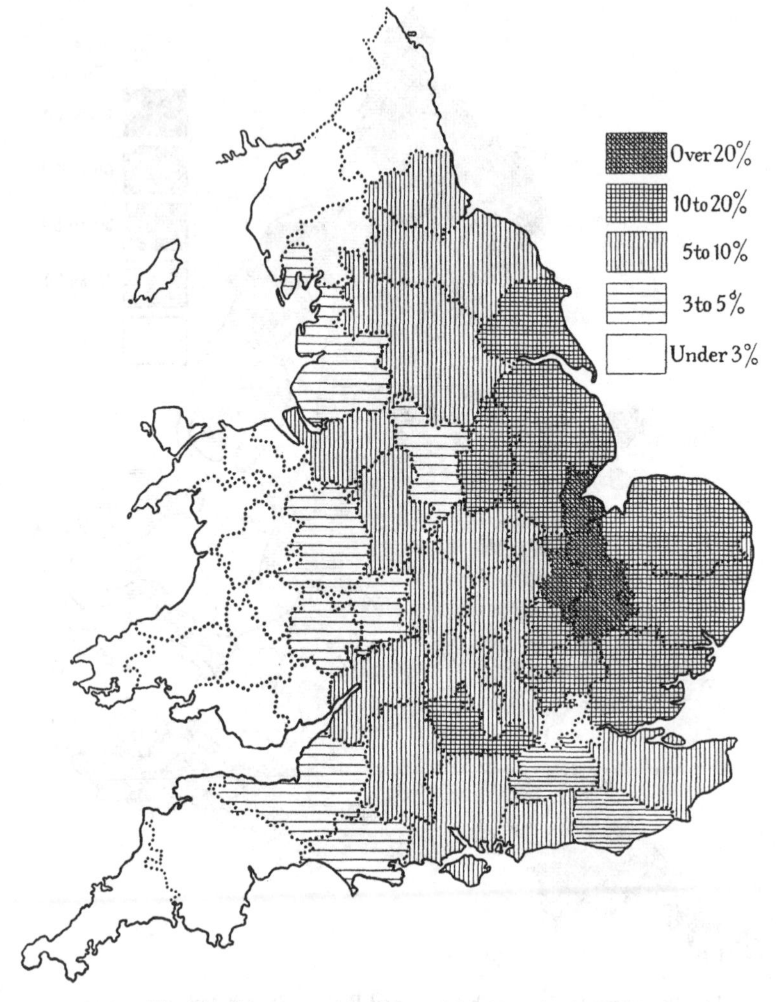

Fig. 19. Percentage of cultivated land in wheat, 1938.
Figures from *Agricultural Statistics*, vol. LXXIII, Pt. 1, 1938.
Acreage and Livestock Returns.

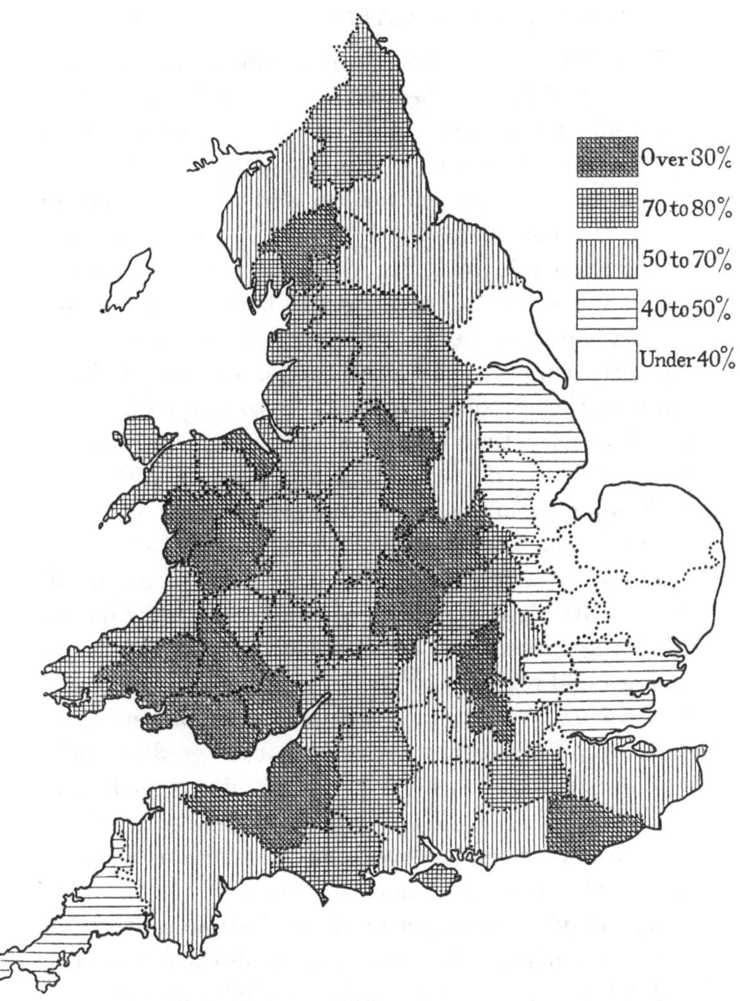

Fig. 20. Percentage of cultivated land in grass, 1938. Figures from *Agricultural Statistics*, Vol. LXXIII, Pt. 1, 1938. Acreage and Livestock Returns.

THE EFFECT OF ALTITUDE AND POSITION

Land lying at the bottom of a slope is moister than land lying higher up, because it receives the water that has drained down from above as well as its own share of the rainfall. Sometimes it is so wet that it forms a marsh unsuited for cultivation and therefore left in grass; the land immediately above is then commonly used for general crops. But where the water-level is well below the surface of the soil the bottom land is not marshy but is on the contrary highly fertile and is more regularly supplied with water than the higher land. Land at the top of a slope may be too exposed or too cold for cultivation, and often, if it lies about 700 ft. in England, it is left either as grass, wood or waste; the limit is higher in places, e.g. the milder parts of Wales, Cornwall, the Yorkshire Wolds, etc.

The small difference in temperature between a north and a south slope may have a considerable effect on the crop, vegetation on a south slope being a little more forward and ready for market before that on the north. In ordinary agricultural practice this is not usually of much importance, but it is for market garden work; considerable differences in price are often attached to small differences in time of marketing.

More important, however, than high mean temperature is the absence of spring or autumn frosts. Low-lying valley lands are peculiarly susceptible to frosts on clear calm nights; the cold air drifts down from above and collects in the valley where it chills the trees and not infrequently kills the fruit blossom and the tender shoots of early potatoes. Land lying above this stagnant pool of cold air escapes these frosts and is therefore a safer place for susceptible crops, even though its mean temperature may be lower than that of the valley. Where, however, the low land adjoins the sea or any great body of water it is protected from these frosts and is, indeed, better than land lying farther off because it is warmer.

We can now understand why fruit is so often grown in undulating country. Slopes are needed to give the desired shelter and aspect, but above all to avoid risks of late frosts. The "lucky banks" of the Evesham district, on which crops can nearly always be raised, are of this character. In the fruit-growing region of Kent the fruit tends to collect on the middle slopes, hops on the lower ground (or wood, if the ground is wet), and woodland or nuts on the higher ground. But near the sea—and this holds generally round the coast—fruit and early potatoes and vegetables can be grown with advantage on the lower ground.

To some extent it is possible to modify the dominating effects of climate on crops; there is a little elasticity in both directions. Plants are somewhat plastic and can be moulded to a certain degree in the plant breeder's hands; they can be bred to tolerate greater cold or more drought than usual. Thus in Canada wheats are being produced to grow farther north than ordinary kinds so as to take into cultivation a belt of land at present suited only for grass. In Australia wheat has been bred to tolerate drought and grow in the drier regions. This sort of work is being done very widely and we do not yet know how far it can be pushed.

Climate apparently cannot be altered; indeed, we can only get over some of its bad effects in one direction. Temperature is at present beyond our control: the hot regions remain hot and the cold regions remain cold and we cannot alter matters. Wet districts can be improved somewhat by drainage. The great triumph, however, is in the dry regions. Irrigation and the special agricultural methods known as dry farming have brought into cultivation enormous tracts of land, hitherto desert, in the United States, India, South Africa, Canada, Australia, the Sudan, and elsewhere. But this modern triumph is no modern invention. Egypt and Mesopotamia were irrigated long before our civilisation appeared.

ECONOMIC FACTORS

Climate and soil determine what crops can be grown, but economic factors (including Government policy) come into play and determine what crops actually shall be grown.

As the produce of the land has to be sold it obviously must be got to the market; chief among the economic factors is therefore the question of transport. From this point of view livestock presents least difficulties; animals can be made to walk to market while crops have to be carried. Sheep, cattle, and horses therefore tend to become the mainstay of agriculture in countries lacking transport facilities. But where transport is better wheat or maize becomes a convenient product, for either will keep almost indefinitely so long as it remains dry, suffering little if any deterioration, however long the journey to market may take, and, what is more im-

Treeless prairie, first stage of development.
Ranching (Sounding Lake, Alta.).

Treeless prairie, last stage of development. An experimental farm (Indian Head, Sask.).

Fig. 21. Development of prairie land, Western Canada.

portant from the farmer's point of view, both are always saleable. On the other hand, fruit and vegetables will not keep and can only be grown where transport facilities are good and refrigeration, if necessary, is available.

History repeats itself with but little variation in the agricultural development of virgin countries in temperate regions. At first the country is pastoral. Then with the opening of railways comes wheat or maize production. Later on when the country is more closely settled other crops are raised and wheat loses its premier position: for a time oats are wanted in enormous quantities for the horses used in railway and other construction work. Then the quality of the livestock has to be improved: fat lambs, good quality beef suitable for chilling, and milk sold on butter-fat content are required, and these need green crops and improved pastures. Green crops are wanted for the cattle and sheep, and other crops are wanted to satisfy the more exacting needs of the population that follows the simple-living pioneer. Then as transport becomes still easier fruit and vegetables are raised in suitable districts for shipment to the cities or abroad (Fig. 21). Canada, South Africa, Australia and the United States show all these stages of development.

CROP PRODUCTION IN GREAT BRITAIN

In Great Britain these various factors have brought about much specialisation in crop production (Fig. 22). The warm districts of Cornwall, the Channel Islands and parts of the Ayrshire coast yield early crops; the dry districts of the eastern counties produce wheat, barley,

Fig. 22. Crop map and isotherms of Great Britain. ("Roots" include sugar beet and mangolds.)

peas, seed and sugar beet; also much milk, as the result of economic factors, although the natural conditions are not always very favourable to grass. The cooler moister regions of Cheshire, the Fens, the Wash, and the Lothians produce potatoes and sugar beet; oats and swedes are raised in moist regions, while in still wetter districts grass and dairying predominate. Wherever the climate is satisfactory and the railway connections good an industry springs up in fruit and vegetables. Finally, the higher or less accessible districts are given up to raising of livestock to be fattened out in those favoured districts where animal food grows more quickly than the animals born on the spot can consume it.

PART II

THE CONTROL OF THE SOIL

CHAPTER V

CULTIVATION

In the preceding chapters we have been dealing with the soil as it stands in the field and studying the changes which it undergoes in the natural state. We now turn to the second part of our subject: the methods whereby the farmer can utilise the soil to the greatest advantage and make it yield crops in quantities that repay the time and trouble involved.

Practical growers have long since discovered that crops will not grow unless the land is properly cultivated.

The first object of cultivation is to get the soil into a good "tilth", i.e. to make it assume the nice crumbly condition which long experience has shown is best suited to the growth of plants. The first stage is to bring to the proper size the lumps or clods in which the soil exists. An important fact is that the clay can exist in two states, the sticky and the crumbly state (p. 30): and it is such a dominating substance that it confers these properties on the soil. In consequence any soil which contains more than about 10 per cent of clay may appear in two very different guises—either in the nice crumbly condition of a fine tilth, or in the other state when it is sticky after a spell of wet weather, and dries into hard clods in dry weather.

The second object of cultivation is to free the land of weeds, which has to be done by uprooting them and leaving them on the surface to perish for want of water, or, in the case of annuals, burying them deeply enough to prevent their springing up. Weed seeds unfortunately cannot all be killed in this way, but must be encouraged to germinate; the young seedlings are then uprooted or buried. This process is called "cleaning" the land.

AUTUMN CULTIVATION

This should begin as soon as possible after the corn is cut, if possible actually during the harvest. If the stubble is fairly clean the land can be ploughed at once. Alternatively, and this is often better on light land, a broadshare can be sent over the land to skim off the surface and leave it to dry. Weeds and grass are thus killed, but the ground below remains moist enough to afford weed seeds a chance of germinating. Winter ploughing can then be begun when convenient.

On heavy lands the earlier this is done the better. The special objects here are to ensure that the land shall dry quickly in spring and that the soil shall take on the proper crumbly condition. The former is secured by leaving the land in a rough state so that a considerable surface is exposed, and also by arranging that water can easily get away. The latter is effected by a combination of good cultivation with the addition if necessary of lime, chalk or limestone, and of organic matter, and by exposure to frost. On the other hand, long exposure to water (e.g. a long wet winter), alkaline substances such as liquid manure, much trampling by horses and men and general bad management, all tend to change the crumbly into the undesirable sticky state.

On light, sandy or peaty land it is not necessary to start so soon: the main object is to kill weeds, and delay is permissible so long as no seeding takes place.

The process of cultivation is carried out first by ploughing, then by harrowing and rolling.

PLOUGHING

The plough consists essentially of two cutting edges, a vertical one called the coulter and a horizontal one called the share, and a curved steel or iron plate called the mould board which turns over the slice of earth as it is cut. The traditional English style of ploughing is to turn the slice without breaking it, so as to form a succession of parallel ribbon-like strips over the whole field.

On light soils the slice is turned right over so that the surface remains pretty level: thus the surface weeds are buried and the seeds below the surface have a chance of germinating. On heavier soils, however, while the vegetation is still ploughed in, the slice is left on edge with high crests so as to expose more surface and to allow more complete penetration of frost and at seeding time a better entrance to the tines of the harrow.

In the middle of last century the mould board was modified by the American makers so as to break the slice as it turned over: these ploughs were subsequently called "digger" ploughs or Oliver ploughs from a successful maker in the eighties. The work done is not so nice looking as in the old English style, but it is probably more effective, excepting in a wet season when the surface is liable to be beaten down too much. These different types of furrow-slices are shown in Fig. 23.

A team of horses can plough about ¾ acre[1] per day on heavy land in England; however urgent the need they can hardly do more. It is obvious therefore that a wet autumn and winter gravely handicaps the farmer and leaves him with his work sadly behindhand in spring. This handicap has now been lessened, and by using a modern tractor it is possible to work to a greater depth than with horses and at two or three times the speed.

It is not desirable to plough to the same depth every year, as the plough sole tends to compact the subsoil and form a hard layer, called the plough pan, which roots cannot penetrate. Periodically the plough should go deeper so as to get underneath this layer; this can well be done for sugar beet, mangolds, or potatoes. In the Rothamsted mangold experiments, where the land had been ploughed deep every alternate year, the depression due to shallow ploughing in one year was only 0·4 ton per acre, but where the land had been shallow ploughed for 2 years or more in succession the depression in yield as compared with continuous deep ploughing was 1·7 tons per acre. Deep tillage with the cultivators is never harmful, and may be very beneficial. In 1935, for

[1] The old standard was an acre a day, and this is the figure given in the old song:

> "We've all ploughed an acre, I'll swear and I'll vow,
> For we're all jolly fellows that follow the plough."

An acre was originally defined as the area that a team could plough in one day and so it varied in different regions.

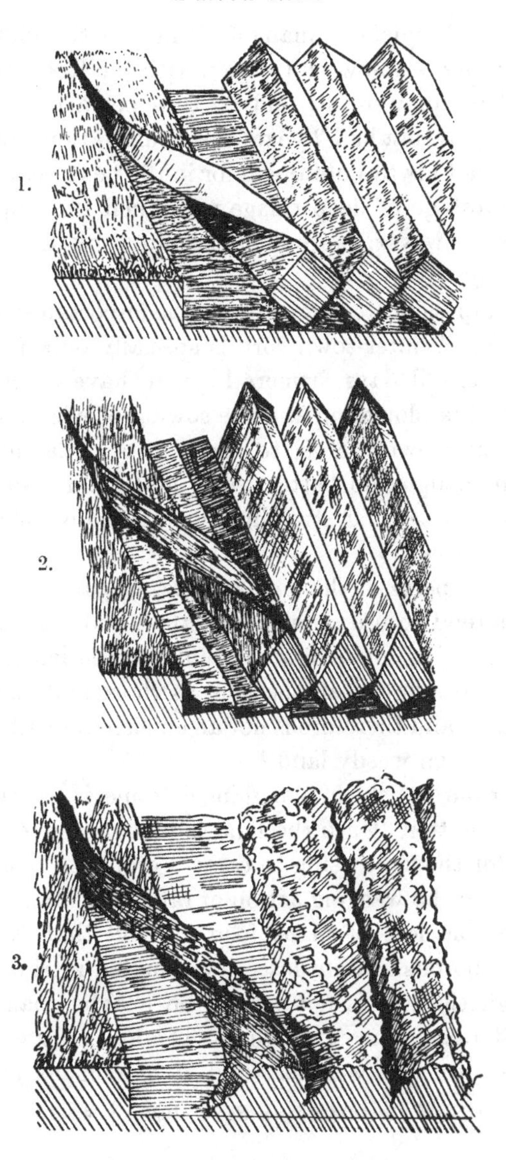

Fig. 23. Types of furrow-slices.
1. Rectangular unbroken. 2. Crested. 3. Rectangular broken.

example, the yield of mangold roots on the plots that had 2 years' shallow cultivation with the Rototiller and the cultivator were 5·1 and 8·7 tons per acre less than on the plots that had 2 years' deep cultivation. But this effect is not always obtained, for in the next year (1936) the advantage of deep tillage was much less apparent. Cereals, however, were not benefited by the deeper ploughing.

It is impossible to overestimate the importance of good ploughing. Winter-sown corn is specially dependent on it. In the old days farmers liked to have an interval between the ploughing and the sowing of winter wheat, or, in their own words, to sow "on a stale furrow". Modern ploughs do better work, and good results are obtained by sowing straightway after ploughing on a "fresh furrow".

In some parts of the country, e.g. in parts of Wales, farmers tend to use a quicker method and scarify their land instead of ploughing it. While the scarifier cultivates it does not clean the land as thoroughly as good ploughing, and therefore is not an efficient substitute for the plough on weedy land.[1]

The time at which ploughing is done is a very important matter. Once the land is ploughed it not only ceases for the time being to carry a crop, but it is more exposed to the washing action of rain, which may mean considerable loss of plant food. A soil that is at all light or porous readily loses its valuable nitrates, and although the loss is not serious on heavier loams and clays, it takes place even there. The loss only goes on, however, in wet weather, and if one could rely on

[1] See *Rothamsted Annual Report*, 1936, pp. 45–8.

having the ground dry or frozen hard throughout the winter it would be simple enough to arrange about cultivation; the soil would be turned up roughly at the beginning of the winter, and left to the end. Unfortunately our winters are variable: land turned up in autumn may only occasionally get frozen, and may lie wet for long periods; it then derives less benefit from the cultivation, and suffers considerable loss of plant food.

In general, light soils may be left uncultivated until late in the winter or early spring, partly because the amount of frost they require is only small, partly because the loss that they would suffer in rain is very large. Heavy soils, however, should be turned up in the autumn as early as possible because they require a large amount of frost, and suffer only little loss. The rule, of course, requires intelligent application and adaptation to each locality.

SPRING CULTIVATION

The object of the spring cultivation is to complete the winter cultivation and obtain a tilth suitable to the crop. The operations depend on the fact that if the soil is in the proper state the small lumps readily fall to pieces when just sufficiently moist. Hence after winter the soil is allowed to dry and is then broken down into small lumps by a clod-crusher: these are then allowed to become moist so that they break to pieces with harrows. The operation probably demands more skill and judgement than any other part of soil management: both the drying and the rewetting are done by the weather and are therefore largely out of the farmer's control. Only a limited time is available and the season for sowing the crop soon passes; on the other hand, if the farmer tries to hurry matters too much and begins before the land is dry enough he may "poach" it so badly as to ruin it for a time. His only chance, therefore, is to have his work well forward in autumn and winter so as to take full advantage of favourable opportunities in spring. The ideal case arises when the soil was turned up roughly in autumn, when the winter was frosty, and the spring just sufficiently showery to soften the lumps at the proper time and facilitate their breaking down under the harrows. A mild winter makes matters more difficult because the clay does not become crumbly. If the early spring is dry the soil dries to a hard surface, or, in the case of lighter soils, to steely lumps which can be broken by a suitable roller; if, however, the spring remains dry there is great difficulty in obtaining a proper tilth; this is often experienced on the loams of the south-eastern parts of England where spring rainfall is low and winters are mild. A good cultivator watching his opportunities often achieves remarkable results, but no one has yet succeeded in reducing his art to an exact science.

The most difficult case arises when the winter is mild and wet and the spring wet so that the soil never dries for the preliminary breaking down into small lumps. The heavier the soil the worse the trouble; and seasons of this kind have sometimes proved disastrous. No way round the difficulty is yet known, and this is one reason why heavy soils are often not brought into arable cultivation.

The disk harrow is very useful in spring in mitigating these effects, but the tilth produced may look better than it really is: "forced" tilths are not much in favour with farmers.

The introduction of the motor on to the farm has opened up new possibilities in the way of cultivation. The usual implements were all developed for the horse, an animal which pulls. But the motor is designed to turn a wheel round, and although this can be translated into a pull, it would be better if the rotating motion could be used direct for cultivation. Rotary implements are now on the market and they have proved very useful for certain types of work, e.g. in orchards and market gardens where little space is available for headlands. In the Rothamsted experiments rotary cultivation has usually been somewhat less effective than the ordinary ploughing, though better than cultivating with the cultivator alone. When working to greater depths all three implements gave about the same result. There are considerable possibilities in rotary cultivation.

ROLLING

Rolling is used for several purposes: for consolidating the seed bed before the seed is sown; for breaking any surface crust formed by rain before the plant is well established; for undoing the effect of frost in lifting plants out of the ground; for destroying weed seedlings appearing during the early spring in winter-sown corn; and for pushing stones below the surface so that they cannot injure the binder blades.

The beneficial effect on the seed bed was shown in the Rothamsted experiments in 1934;[1] the increased yield of sugar-beet roots given by sulphate of ammonia was

[1] *Rothamsted Annual Report*, 1936, pp. 39–40.

2·26 tons per acre with heavy rolling, and 0·96 with ordinary rolling; the corresponding figures for the total sugar were 6·8 and 1·9 cwt. per acre respectively. Beneficial effects, however, are not always obtained; the 1935 experiment, in which agricultural salt was used, showed no such results.

Another example of the beneficial effect of consolidation is seen in the extra growth of wheat and barley occasionally seen where the wheels of the tractor have passed.

After the young plant is up great judgement is needed to decide whether rolling or harrowing is the better operation to perform; a mistake in either direction may cause much harm.

At Rothamsted rolling and harrowing each gave definite but small increases in yield of wheat, and the two together gave a larger increase: the gains were, in cwt. per acre:[1]

	Rolling	Harrowing	Rolling and harrowing
Grain	0·5	1·0	1·7
Straw	2·4	-0·5	-0·6

In the Cambridge experiments[2] rolling and harrowing were without effect on yield except on light chalky soils, where rolling was very beneficial but harrowing was not (Table VI).

Table VI. *Effect of rolling and harrowing on wheat growing on light chalk soil*, 1934–5

	Bushels per acre
No spring cultivation	17·3
Rolled only	21·2
Harrowed only	19·2
Rolled and harrowed	23·5

[1] *Rothamsted Annual Report*, 1936, p. 40.

[2] H. G. Sanders, *J. Farmers' Club*, 1935, p. 87.

The production of a fine seed bed necessitates the use of implements which give the greatest amount of shattering of the clods; the binding of the soil together requires other implements and other conditions of work.

The effect of these different implements in breaking up the soil clods is estimated at Rothamsted as follows: four 16 in. sieves are used, fitted with bottoms having apertures $1\frac{1}{2}$, $\frac{5}{8}$, $\frac{1}{4}$ and $\frac{1}{8}$ in. respectively. A spadeful of soil is taken, put on the top sieve, and the bank is carefully shaken. The soil on each sieve is weighed and the results are then expressed as the percentage of the sum of the weights found.

SUMMER CULTIVATION

Summer cultivation consists in hoeing and its object is twofold: to keep the surface of the soil in a fine state and to kill weeds. A layer or "mulch" of fine soil keeps the land somewhat cooler during hot weather and to this extent prevents loss of moisture. This was shown many years ago by the American investigators, Kedzie of Michigan and King of Wisconsin.[1] In dry climates the gains in soil moisture are pronounced: surface cultivation with disk harrows is started directly a shower of rain has fallen, and before the land has had time to dry up again. This is the central idea of "dry farming". Unfortunately it has led to much soil erosion.

The effect of hoeing on soil temperature and soil moisture is not very marked in our climate, though it can be demonstrated on a hot sunny day; not usually,

[1] See Kedzie, in *Michigan Agric. Coll. Rep.* 1889, and F. H. King, *Wisconsin Agric. Exp. Station Rep.* 1889.

however, on a cold one. It can be done by the following experiments:

Set out three plots each 3 × 3 ft.: leave one alone entirely so that it may cover itself with weeds; keep the second clean by hand weeding but do not touch it otherwise; keep the third well hoed. Take readings of the temperature of the soils at depths of ½, 3 and 6 in. below the surface; in particular obtain readings on hot sunny days and on cold sunless days. Periodically also take samples for the determination of the moisture content: this is done by driving in the soil borer to a depth of 6 in., weighing the soil that is withdrawn (or a fair sample of it), leaving to dry in a warm place and then weighing again. Table VII gives some of the results which have been obtained.

Table VII. *Effect of hoeing on moisture content and temperature of soil*

	Hot sunny day ° C.		Cold sunless day ° C.	
	Untouched	Hoed	Untouched	Hoed
Soil temperature, ½ in.	35·0	31·5	17·5	17·0
,, 3 in.	30·5	28·8	16·7	16·3
,, ,, 6 in.	27·0	26·5	15·8	15·5
Soil moisture, per cent	14·7	16·0	19·3	18·4

The effect in saving water is too small to be worth having under English farm conditions; indeed damage to the plant roots may cause summer cultivation to be harmful. Experiments at Rothamsted[1] showed that inter-row cultivation of sugar beet and of kale not only did no good but actually reduced the yield, and in no case has summer cultivation been of value except where it was needed to keep down weeds.

[1] *Rothamsted Annual Report*, 1936, pp. 37–48.

Weed destruction, indeed, appears to be the chief effect of cultivation after the crop is sown; any other equally good method of killing weeds might be expected to have the same effect on crop yield. In Belgium and Holland spraying with cyanamide or kainit is used for killing weeds, and is both cheaper and better for crop production than cultivation. Weeds must, however, be destroyed, as they reduce crop yields more than anything else: even more than starvation of the land.

EARTHING UP

Some crops are benefited by drawing the earth up to them so as to form a ridge. Potatoes give a larger proportion of saleable tubers; special implements have therefore been devised to mould them up. The effect is complex: the process affords a thorough hoeing, it facilitates drainage and the drying of the soil: indeed it may make the soil too dry.

RIDGING

In the northern counties and in Scotland it is customary to lay the land in ridges on which turnips or swedes are grown. The process facilitates drainage, evaporation and weed destruction. In the south, where there is usually too little rather than too much water, ridging is adopted only for the mangold crop.

FALLOWING

Fallowing, or leaving the ground free of crops, gives the farmer a free hand for his cultivations, and much increases the stock of nitrate in the soil: the growing plant apparently influences bacterial activity and so reacts on nitrate accumulation (Fig. 24). There is more nitrate present on fallow land than on cropped land even after allowing for what has been taken up by the crop (Table VIII).

Fig. 24. Showing the difference between mustard grown after a clover-rye-grass mixture (left), and after fallow (right) (Rothamsted).

Table VIII. *Nitric nitrogen present in soil and in crop on fallowed and on cropped soils respectively*

	N as nitrate lb. per acre			
	June 1911		June 1912	
	Fallow	Wheat	Fallow	Wheat
In soil	54	15	46	13
In crop	—	23	—	6
Total	54	38	46	19
Deficit in cropped land	—	16	—	27

Wherever the climate allows, it is good practice to plough very early in autumn and cultivate well so as to kill weeds and to give the bacteria a good chance of producing nitrates for a winter-sown crop. This is called a bastard fallow. It may even be profitable to secure an earlier start by sacrificing the aftermath of the seeds ley.

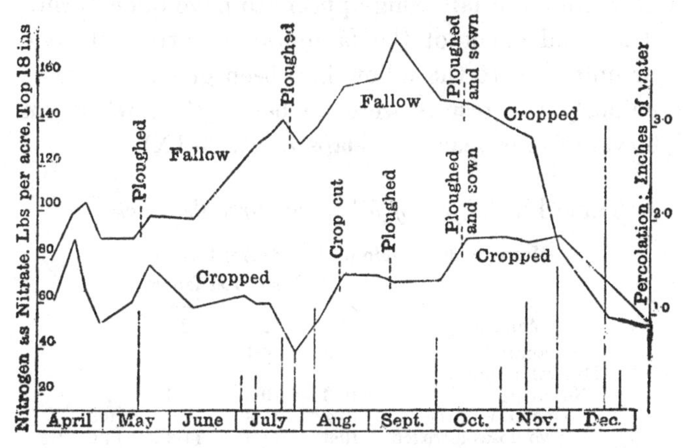

Fig. 25. Effect of fallowing on nitrate content of soil. The vertical lines show the amount of water percolating through the soil. From May onwards to September the fallow land rapidly gains nitrate: then the loss begins and goes on rapidly throughout November and December when percolation is high. The changes are less on cropped land: a rise in nitrate content begins in October after the first ploughing but it soon comes to end after the new crop is sown.

On heavy land it used to be worth while to spend a whole season over the fallow, sacrificing rent, rates and capital charges, so as to allow ample opportunity for obtaining these various effects of spring and summer cultivation (Figs. 24 and 25). Remarkable increases of yield have been obtained at Rothamsted as the result of fallowing though the effect has not been perceptible for more than one year. A strip of the Broadbalk wheat field is fallowed each year: here the crop grown immediately after the fallow is double that on adjacent land which had been fallowed two years previously. The differences are less marked where nitrogenous fertiliser is given, and in the plot receiving the treble dose of nitrogen fallowing appears to have done harm.

The good effect of the fallow apparently lasts one year only: where the fallow had been given two years previously the results were no better than where an interval of four years had elapsed (Table IX).

Table IX. *Effect of fallow on Broadbalk wheat*

Mean yields of grain (cwt. per acre) 1935–9

	Year after fallow			
Manuring	1	2	3	4
No manure	12·5	6·4	7·1	7·1
Mineral manure:				
No nitrogen	16·3	9·5	7·9	7·5
+ One dose nitrogen	19·4	11·9	10·0	10·9
+ Two doses nitrogen	19·9	15·8	14·2	14·0
+ Three doses nitrogen	20·7	18·5	15·8	14·9
Farmyard manure	20·9	16·6	15·9	14·3

SUBSOILING AND TRENCHING

The object of these operations is to increase the root range of the plants.

In ordinary circumstances plants do not have a great deal of root room; the surface layer, which is best suited to their requirements, is only about 6 or 8 in. deep—not always as much, indeed—and it is usually underlain by a subsoil which is less suited to the plant and from which its roots cannot draw much nourishment. Any process that makes the subsoil a better habitat for the roots increases the extent of the root range and therefore enables the plant to make better growth.

The improvement of the subsoil is not commonly attempted in farm practice excepting only on arable soils where a plough-sole or a pan occurs near the surface; recourse is then had to deep ploughing or subsoiling. The operation is not necessary oftener than once in four or five years, and it is usually done as part of the preparation for the potato or sugar beet crop. In spite of much dogmatic assertion there is no good evidence that subsoiling is of value except where a plough-sole has to be broken. In the Rothamsted experiments subsoiling failed to increase the yield of sugar beets.

In market gardening and horticulture it is common to trench the land with the object of making the subsoil more like the surface soil. Years ago it was thought that the subsoil was really the virgin soil, rich in stores of food that only needed liberating by the action of frost. Experiments have shown that this is not correct; the subsoil is really very poor in plant nutrients, and nothing whatever is gained by bringing it to the surface. Considered as a manure it is despicably poor. This is the general rule; exceptions arise when the subsoil contains much chalk or marl, and the surface soil does not; or when the subsoil is clay, and the surface soil is fen or too light a sand. With these exceptions the subsoil is much poorer than the surface soil, and therefore to make it equal the gardener must add manure to it.

To get the subsoil into the same mechanical condition as the surface soil without bringing it up to the top is not easy because frost does not penetrate readily. Something can be done, however, by adding lime, limestone, chalk, or basic slag, to the subsoil at the time of trenching.

The roots of the plants have a wonderful facility for boring their way into the subsoil, and very stout roots can often be found well below the surface depth. It is not clear, however, that the loosening of the soil is particularly helpful to these plants, indeed a soil which is simply loosened and then left soon settles back to its natural condition.

Three methods of trenching have been used:

(1) The top spit is kept on the top, and manure is buried in with the subsoil.

(2) The digging is done in the same way but no manure is added, the subsoil being simply loosened.

(3) The subsoil is put on the top and the surface soil below.

These three methods have given rise to much discussion but there are times when at least two of them are sound.

The first is practically always beneficial, though it is not always a commercial success.

Experiments at Woburn Fruit Farm and at Rothamsted have shown that the second method (the loosening of the subsoil without additional manure) has very little effect either on the water content, the amount of plant food or the growth of fruit trees. There is no evidence that this operation is worth doing; the gardener who takes the trouble to trench should certainly not miss the excellent opportunity it affords for putting the very necessary manure into the lower spit.

There are, however, cases where the third method (the inversion of the surface soil and bringing up of the subsoil) has worked well, particularly on sandy soils where the difference between the surface and the subsoil is less than on the loams and clays. The subsoil is not particularly unsuited for the growth of plants, and when brought to the surface it only requires proper manuring to enable plants to make a satisfactory start. Then when the roots grow down to the second spit they come to the old surface soil and develop well: thus in the end they range over two spits whereas on untrenched land they cover one spit only.

GRASSLAND CULTIVATION

Proper cultivation is one of the secrets of the successful management of grass land. During wet weather the soil may be so compacted by the stock, that the grass and clover roots cannot obtain enough air. Cultivation with a deep working harrow is then very necessary. Special implements, such as the pitch-pole harrow and the rejuvenator, have been devised for this purpose, and they not infrequently bring about remarkable improvement.

On light chalky soils, however, the soil may become too light. Earthworms burrow in the soil and throw out material from below on to the surface. These casts accumulate steadily at a rate calculated by Darwin to be about 1 in. in 10 years. The worms honeycomb the ground with their burrows, and this action, though necessary to the plant, becomes after a certain stage harmful and injures the grass while allowing moss to grow. The ground should be rolled during dry weather in spring until the proper degree of compactness is reached.

SUMMARY

We can now make a general summary of the chief effects of cultivation; they are:

(1) To help build up the soil particles into the crumbs that give the good tilth: and as part of this action to change the clay—and therefore the soil—from the sticky state which is bad for plants to the useful crumbly state.

(2) To give the crop a clear field for growth and reduce competition by weeds.

(3) In horticultural and market garden practice to change the subsoil and make it more nearly like the surface soil.

CHAPTER VI

THE CONTROL OF SOIL FERTILITY

We now turn to the final part of our subject: the study of the methods by which the fertility of the soil may be increased, and the soil may be made more favourable for the growth of plants. The plant requires from the soil six conditions, viz.:

1. Proper water supply.
2. Proper air supply.
3. Suitable temperature.
4. Nutrient salts.
5. Ample root room.
6. Absence of injurious substances or pests.

These six are all quite distinct: it is insufficient to satisfy five of them if the sixth is not attended to: any single one left unsatisfied may operate as a limiting factor and render the soil infertile.

Thus the problem of increasing the fertility of the soil reduces itself to the discovery, first of the factor or factors limiting the growth of the crop, and then of the best methods of overcoming the limiting factors.

Water supply. The water supply comes from the rain, but if the soil lies on the lower part of a slope it may derive further quantities from the land higher up. Crops vary greatly in their need for water, according to the conditions. In hot dry climates from about 300 to about 900 lb. of water may be transpired by the plant for each lb. of dry matter made. The water requirements per lb.

of dry matter of the common crops in these circumstances are as follows:

Most economical: (300 lb.)	Millet, Maize.
Less economical: (400–600 lb.)	Sugar beet, barley, oats, potatoes, cowpeas.
Least economical: (800–900 lb.)	Lucerne, clover, grass, flax.

In cooler moister climates, as in England, the figures are smaller and the differences between the crops are less. Wheat appears to require only about 250 lb. of water per lb. of dry matter produced: for a 40 bushel crop this means some 600 to 700 tons of water per acre, equal to 6 to 7 in. of rain. A 30 ton mangold crop requires about 12 to 14 in. of rain. The need for water is greatest when the plant is actively growing, and in the eastern parts of England the rainfall may be insufficient to satisfy the crop. The reserves of water in the soil then become very important.

The plant needs water least when it is making seed: the best seed comes from the dry eastern parts of England.

In Great Britain it is not usually practicable to add water except in market gardens where spray irrigation is used, and in some meadows where flooding is possible. In dry countries irrigation is usual, but it needs to be done with great care or much harm results. Control of the water supply is effected in farm practice by adding farmyard manure, thereby increasing the power of the soil to hold moisture (p. 241), and by bringing the soil into a good tilth.

Soils vary greatly in their power of holding water and

of giving it up to plants. Sandy soils can hold only little but they give up almost all they have; loams hold more but do not give it up quite so completely; clays hold a good deal more but also retain much more (Table X). Thus the supply available to the plant varies much less than the water-holding power of the soil.

Table X. *Water-holding power of soils per cent*[1]

Soil	Held so that it will not drain away ("moisture equivalent")	Held so firmly that the plant cannot get it ("wilting coefficient")	Difference, available to the plant
Light sand	3·3	1·3	2
Sand loam	10–14	3–5	8
Silty loam	17–25	7–10	10–15
Clay	29	20	9

AIR SUPPLY

This is closely bound up with the water supply because the only space available for air is that part of the pore space that is not filled with water. The air supply is thus determined by tilth and drainage.

Some plants need much more air than others for their roots: maize and peas can hardly get enough, while rice and some of the *Salix* family can do with very little. Good aeration favours the development of a large fibrous root system, vigorous growth of large green leaves, and good setting and developing of fruit. Root respiration is closely connected with the uptake of plant nutrients.

[1] F. J. Veihmeyer, noted in J. V. Botelho da Costa, *J. Agric. Sci.* 1938, **28**, 660.

TEMPERATURE

This also is largely a question of water supply, since a given amount of sunshine will raise a dry soil to a higher temperature than a wet one. But it depends also on the slope of the land, a south slope being warmer than a north slope; also on whether the soil is bare or shielded from the sun by a cover of vegetation.

ROOT ROOM

The depth of the soil may be limited by a high water-table, a pan or layer of rock, or by the solid rock.

INJURIOUS SUBSTANCES

The commonest are the soil acids, either organic or inorganic; but injurious substances are also formed in waterlogged soils.

SOIL TYPE

The effects of these various conditions in different soils under similar weather conditions are shown by the following experiment due to S. T. Parkinson.[1] A trench was dug 5 ft. broad, 30 ft. long and 3 ft. deep: the sides and bottom were lined with loose bricks and stones, and four partitions were put up. The divisions were then filled respectively with a good loam, a peat, a gault clay, a poor sand and broken chalk. Carrots were sown and gave results shown in Fig. 26.

The loam gives by far the best results. The small growth on the sand is due to lack of nutrient salts: that on the chalk both to poverty and to a direct harmful

[1] *J. South-Eastern Agric. Coll. Wye*, 1910, pp. 258–61.

effect of the chalk itself. The stunted roots on the peat are attributable to the acidity, and the fanged roots on the gault to lack of air supply, possibly also to some harmful substance: acidity has this effect.

Soil management would be greatly facilitated if it were easy to convert all soil types to loams: to add enough sand to clay soils, or enough clay to sandy soils,

Fig. 26. Carrots grown on various types of soil. (S. T. Parkinson's experiment.)

to effect the transformation. Unfortunately this is not usually practicable. Clay soils would require 100 tons or more of sand per acre to make an appreciable difference, which would cost too much. Clay can be added to a sandy soil at less expense because a dressing goes farther than in the case of sand: the operation used to be fairly common using a clay containing calcium carbonate,[1] so that two desirable constituents were added in one operation.

[1] This mixture is called marl.

Looking down the list on p. 109 it is clear that good tilth, the deepening of the soil where necessary, drainage, neutralising soil acidity and manuring, are the best ways of raising soil fertility.

TILTH

The methods of obtaining a good tilth are: (1) suitable cultivation; (2) addition of organic matter in the form of farmyard manure or crop residues; (3) where necessary addition of lime or chalk. Heavy clay soils are difficult to bring to a good tilth and hence are unsuitable for cultivation; light soils easily acquire the proper tilth: in between come a whole series of soils which a skilful farmer can manage while an unskilful one cannot.

DEPTH OF SOIL

In some cases a pan or thin layer of rock separates an upper from a lower soil, and once it is broken the roots are able to penetrate to the lower depths. Various mechanical devices have been used for the purpose, particularly in the reclamation of peat soils.

Sometimes, however, the rock is solid, and then it obviously cannot be removed. If it lies in regular layers end on to the surface there is the possibility that some of the roots may be able to find a way in between, as happens in the Upper Greensand beds of West Sussex, but if the layers lie horizontally the chance of success is much smaller. The problem becomes still more difficult when the soil lies on gravel. The "shrave" of West Sussex, and the commons of Hertfordshire, are formed of thin soils lying on gravel which could never be managed

Fig. 27. Harpenden Common. Land that cannot be cultivated because it is a thin soil lying on gravel.

Chalk subsoil. This land can be cultivated although the soil is thin. (Harpenden.)

Gravel subsoil. This land cannot be cultivated because the soil is too thin for a gravel subsoil. (No Man's Land, Wheathampstead.)

Fig. 28. Influence of the subsoil.

economically in spite of the fact that good farmers have always been found on deeper soils round about them. Modern science has as yet no solution to offer (Figs. 27, 28).

DRAINAGE

Waterlogging of the soil has two disadvantages. It cuts off the air supply to the plant roots and therefore prevents the growth of many plants, including some of the most valuable arable crops, grasses and clovers, while allowing docks, thistles, bent grass, buttercups and other weeds to flourish. Waterlogging also leads to reduction of nitrates and other oxidised substances with formation of compounds harmful to plant growth. Ferric compounds are reduced to ferrous, manganic to manganous, sulphates to sulphides, etc. The difference in the iron compounds is seen by the change in colour, greenish or bluish being associated with the ferrous state and red or brown with the ferric. A change can also be seen in the case of manganese: the lower oxide is colourless and the higher ones brown.

Bad drainage is one of the common causes of infertility on heavy soils in this country. It was dealt with in the old days by laying up the land in high ridges several feet wide (often a rod wide) which were commonly not quite straight but curved at each end like a long-drawn-out S, the result of a difficulty in turning the ploughs in the days when a long team of oxen was used. The scheme had the drawback that the furrows were usually too wet and too much on the subsoil for satisfactory growth, and sometimes the plants failed altogether. Even a shallow furrow has a bad effect. Finally the advent of the binder necessitated the use of fairly level ground and made the old high ridges impossible. The provision of furrows for taking away water is still necessary on heavy soils, however, and the land is laid up in "stetches" for this purpose.

In market gardening it is often necessary to raise beds for the growth of crops, especially in wet districts.

After 1823, when James Smith of Deanston, Perthshire, began to draw attention to drainage, large areas of land were pipe-drained. The cost was high and in many cases the result must have involved financial loss, although the contingent benefits in the countryside were probably worth it. There was a long dispute as to how deep drains should be laid: Smith laid shallow drains, and Josiah Parkes, a famous engineer who drained Chat Moss and other great areas, laid deep drains. It appeared afterwards that, as often happens in agricultural discussions, both sides had a good case: shallow drains were needed when the water to be removed comes from above—e.g. from excessive rain or seepage from high land—and deep drains when the water is thrown up from below. Before deciding on the depth of the drains, therefore, it is necessary to ascertain where the water is coming from and how and where it can best be intercepted.

On clay lands the water usually comes as rain and therefore shallow drains are best. The pipes used before 1914 were commonly 3 in. diameter and often laid $2\frac{1}{2}$–$3\frac{1}{2}$ ft. deep and at distances of 15–30 ft. apart, but a well-thought-out plan was always wanted. The cost used to be about £7 per acre. Many of the old drains still remain and would function satisfactorily if the ditches were cleaned out and the outlets put right.

A much cheaper method, however, is now used for land free from large stones: the drainage is done with a mole plough. This implement cuts out a 2–3 in. tunnel 18 in. to $2\frac{1}{2}$ ft. below the surface of the soil into which the water can drain. The tunnel is more permanent than

Fig. 29. The mole plough in operation. Its working part is a steel cylinder with a pointed end carried parallel to the beam which forces a passage through the soil as it goes.

might be anticipated, and may last 15–20 years or more, especially if it does not run straight into the ditch but into the old mains, or, if these cannot be found and cleared, into new drains discharging into the ditch. The work can be done at almost any time of the year, but preferably in spring or summer, and not in very wet or very dry weather (Fig. 29).

Where the land is old grass laid up in ridge and furrow it is often an advantage to run a shallow mole along the furrow.[1]

In Holland and Germany some remarkable draining machines are in use for the extensive reclamations of peat soils there carried out. Some of these make a tunnel like a mole plough and then proceed to line it with a pipe made of wood or other material as part of the work of the machine.

Drainage is a matter for the whole district and not merely for the individual farmer. The ditches must be kept clean for their whole length and the outfalls of the drains open: the main drainage brooks must also be cleaned regularly.

Land that is not wet enough to need actual drains may still require a water furrow to carry away excess of rain; the great point is that water must not stand about on the surface.

NEUTRALISING SOIL ACIDITY

The degree or intensity of acidity is a very important factor in the relation of acid soils to the growing plant. It is measured by a scale called *p*H. On this scale 7 indicates neutrality, lower numbers indicate acidity and higher numbers alkalinity. The scale, however, is not like a foot rule where a difference of 1 in. means the same thing all along its length; it is of the kind called logarithmic and the differences become greater as one moves farther from the neutral point 7. Thus the

[1] See H. B. Turner, *J. Farmers' Club*, Feb. 1926, for details and costs.

difference from the beginning of the scale:

	*p*H	7	6	5	4

is not, as on a foot rule, divided into $\frac{1}{10}$ in.:

		10	20	30
	i.e.	10	10+10	10+10+10
but		10	100	1000
	i.e.	10	10×10	10×10×10

The reason is that when the soil is nearly neutral small differences in acidity are far more important to the growth of the plant than much larger differences when the soil is very acid, and therefore a scale on which the values were equal would not be very useful, nor indeed would it be practicable owing to the wide range of the values.

An approximate idea of the *p*H of the soil can be readily obtained by using one of the reliable soil testing outfits on the market. In this country it is rare to find soils of *p*H greater than 8·5; on the other hand farm soils may go down as low as 4·5 and in exceptional cases lower, and moorland soils may fall to 3·5 or less.

Plants differ greatly in their tolerance of acidity. Some, like Yorkshire fog, sorrel, and rhubarb, will tolerate a great deal: others like rye, potatoes, oats and alsike clover, are not quite so tolerant, while sugar beet, barley, and lucerne do not like it at all (Figs. 30, 31). The tolerance of acidity depends on a variety of factors: e.g. a given degree of acidity is less harmful in moist than in dry conditions, and less harmful in presence of organic matter. In consequence no rigid list can be drawn up, and no two lists agree, but the following is

Fig. 30. Field peas failing on acid soil, Tunstall, East Suffolk. In the background the land has been chalked and carries an excellent crop.

Fig. 31. Sugar beet failing on acid soil at Tunstall, East Suffolk, but growing on the chalked part in the background. Rye, on the other hand, is growing quite well on the acid soil though its stems are a little darker on the chalked part.

roughly the order in which crops usually go in England, those at the top of the list being on the whole more sensitive than those lower down:

Less tolerant	More tolerant
Lucerne and sainfoin	Cabbage and kale
Foxtail (*Alopecurus pratensis*)	Cocksfoot
Barley	Wild white clover
Sugar beet	Swedes
Beans and vetches	Peas
Wheat	Oats
Mangolds	Alsike clover
Red clover	Potatoes
Mustard	Buckwheat
Rye grass	Rye
	Lupins
	Sweet vernal grass (*Anthoxanthum odoratum*)
	Sheep's fescue
	Yorkshire fog (*Holcus lanatus*)
	Sorrel (*Rumex acetosa*)
	Knawel (*Scleranthus annuus*)
	Spurrey
	Rhubarb (most tolerant)

The plants at the top and the bottom of the list can be used as indicators of the state of acidity of the soil. Failure of lucerne, sainfoin, barley or sugar beet, especially in patches, is a fairly sure sign of acidity, as also is the prevalence of sheep's sorrel, knawel and spurrey: the student should learn to recognise these.

Bracken, Yorkshire fog and mayweed are uncertain as indicators; all may occur on chalk soils, and indeed one species of mayweed is widely prevalent on them.

A *p*H of 6·5–7 is usually satisfactory for agricultural crops; values below 6 are undesirable, particularly on light dry soils, except for special reasons.

Crops grown in soils that are too acid show certain

characteristic appearances. The extreme case where the plant only just survives is well seen on the Woburn barley plots that have become acid (*p*H 4·5). The roots are yellowish brown and very stunted with short branches and abrupt ends; they lack the clean white rootlets of normal plants; they do not strike down into

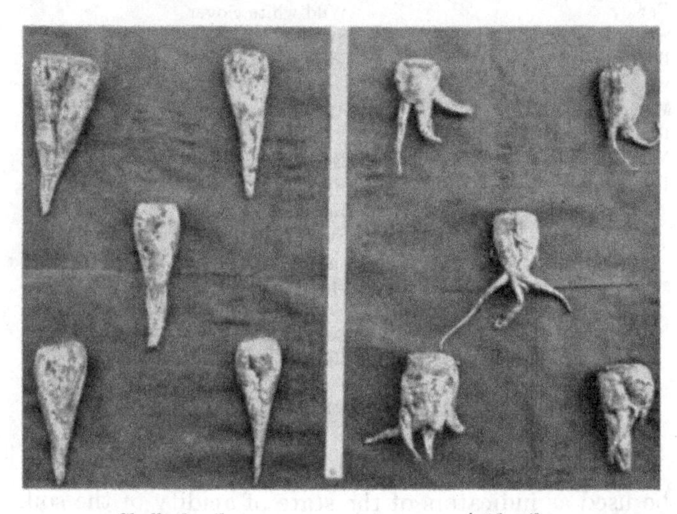

Chalked soil Acid soil

Fig. 32. Sugar-beet roots on experimental plots at Tunstall, East Suffolk, showing the fangs produced on the acid soil and the well-shaped roots that grow after chalking.

the soil but may curl round as if to avoid it. The seedlings develop; the leaves have a dull purplish appearance; they are stiff but the stem never elongates, and at harvest time the plants are only a few inches high, yet they have formed some grain and its nitrogen content is quite normal.[1] Mangolds in like manner develop but

[1] For a full description with chemical analyses see H. H. Mann, *J. Agric. Sci.* 1937, 27, 108–22.

never grow large. Sugar beet take on a peculiar fanged appearance (Fig. 32).

These cases are rather extreme. It is much more usual to find soils that are near but not quite beyond the limit. Their distribution in the field is patchy and this gives a patchy appearance to the crop. Arable crops, especially clover and sugar beet, fail in places and elsewhere thrive vigorously, suggesting a stimulating effect of the marginal degree of acidity.[1] On grass land some species, notably *Festuca ovina* and *Anthoxanthum*, are dark green while others are yellowish.[2]

An important effect of the weakening of the crops on acid land is to allow the growth of weeds which tolerate the acidity but would not tolerate the competition of a full crop.

SOIL ACIDITY AND PLANT DISEASES

The soil micro-organisms are also affected by the reaction of the soil, some being much less tolerant of acidity than others. Disease organisms also are affected: indeed, this affords sometimes the easiest way of controlling them. The organism causing finger and toe in swedes, turnips and cabbages, *Plasmodiophora Brassicae*, tolerates acidity better than its host plants; it is therefore injurious on acid soils. Addition of lime improves the soil for the crop but makes it less suited to the

[1] The permanent wheat on the acid plots at Woburn affords a good example: see E. J. Russell and J. A. Voelcker, *Fifty Years of Field Experiments*, p. 226.

[2] The most striking colour change is that produced in the flowers of hydrangeas. Atkins (*Sci. Proc. R. Dublin Soc.* 1923, **17**, 201) states that the flowers are blue when the soil pH is 6 or less; pink when it is 7·5 or more; and blue and pink on the same plant when the pH is between 6 and 7·5.

parasite. On the other hand, *Actinomyces chromogenus* (*Oospora*), which causes potato scab, is less tolerant of acidity than its host, and over a certain range potatoes can be grown well without fear of attack. Even though their soils need lime for other crops, Cheshire farmers do not add it until after the potatoes are lifted, otherwise "scab" results. The fungus causing "take-all" in wheat (*Ophiobolus graminis*) develops much more readily in slightly alkaline than in acid conditions: it is liable to do damage on calcareous soils but not on slightly acid sands or heavy loams (Fig. 36, p. 141). "Deficiency" diseases are also affected by soil reaction: the "grey fleck" disease of oats due to lack of manganese is less common in acid than in neutral conditions. This is shown in Table XI. While cultivated crops generally require soil reaction nearly neutral, certain plants require acid conditions probably because they cannot tolerate much calcium, e.g. lupins and the heaths.

Table XI. *p*H *values of soils liable to disease, and of soils escaping disease*

Disease	Diseases prevalent	Little or no disease
Finger and toe, Rothamsted	5·85	7·9
Potato scab, Pusey	7·4	6·1
Take-all	6·0	5·0
Grey fleck of oats	6·5 to 7·5	Below 6·0

LIME REQUIREMENT

From the foregoing pages it is clear that there is no such thing as an ideal *p*H for a soil: for some crops and in some conditions one value may be desirable, but for other crops some other value may be better. Unfortunately the *p*H can be changed easily only in one direction; it can be raised by adding lime or calcium carbonate, but it cannot easily be lowered. Addition of acid would of course be effective, but this is not practicable on the large scale. Addition of sulphur or sulphate of ammonia in suitable amounts slowly lowers the *p*H and one or other is used where the change is attempted, as for the prevention of scab in potatoes or of "Take all" in wheat in Great Britain, or in the treatment of alkali soils in dry continental regions; very little, however, can actually be done.

Fortunately the other and easier change is far more frequently required and methods have been devised for estimating how much lime or calcium carbonate must be added per acre to raise the *p*H value by any desired amount; these are described in the books on practical analysis listed in the appendix. A knowledge of the *p*H does not itself show how much lime should be added to make the soil neutral, though it might afford some idea to an expert adviser with local knowledge.

The lime and acidity determinations are perhaps the most generally useful of soil analyses.

STAGES OF SOIL IMPROVEMENT

The history of agriculture affords many instances of the way in which the yield is pushed up by steadily removing the limiting factors that had been keeping it down. From the medieval writers we may infer that 10 bushels was a common crop of wheat in their day: this was obtained on unenclosed land by the use of such stable manure as was available. Later on, when the land was enclosed, it could be kept cleaner: competition of weeds was therefore reduced; still later, rotations were gradually introduced; liming and chalking were more carefully done; drainage was attended to; in the middle of the nineteenth century artificial manures were

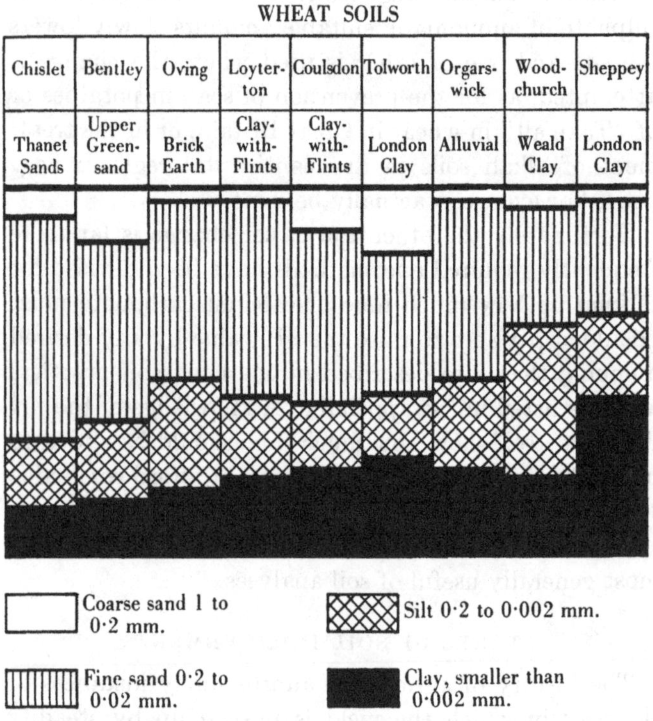

WHEAT SOILS

Chislet	Bentley	Oving	Loyterton	Coulsdon	Tolworth	Orgarswick	Woodchurch	Sheppey
Thanet Sands	Upper Greensand	Brick Earth	Clay-with-Flints	Clay-with-Flints	London Clay	Alluvial	Weald Clay	London Clay

Coarse sand 1 to 0·2 mm.
Silt 0·2 to 0·002 mm.
Fine sand 0·2 to 0·02 mm.
Clay, smaller than 0·002 mm.

Fig. 33. Mechanical composition of soils well adapted for wheat.

introduced and tillages were improved; still more recently improved varieties and better seed have been available so that now 40 bushels are readily obtained by good farmers.[1] Each improvement has consisted in

[1] Mr Alfred Amos obtained 96 bushels per acre in 1918. See *J. Bd Agric.* 1919, **25**, 1161.

removing some factor that was keeping down the yield to a certain level. But there still remain two sets of

POTATO SOILS

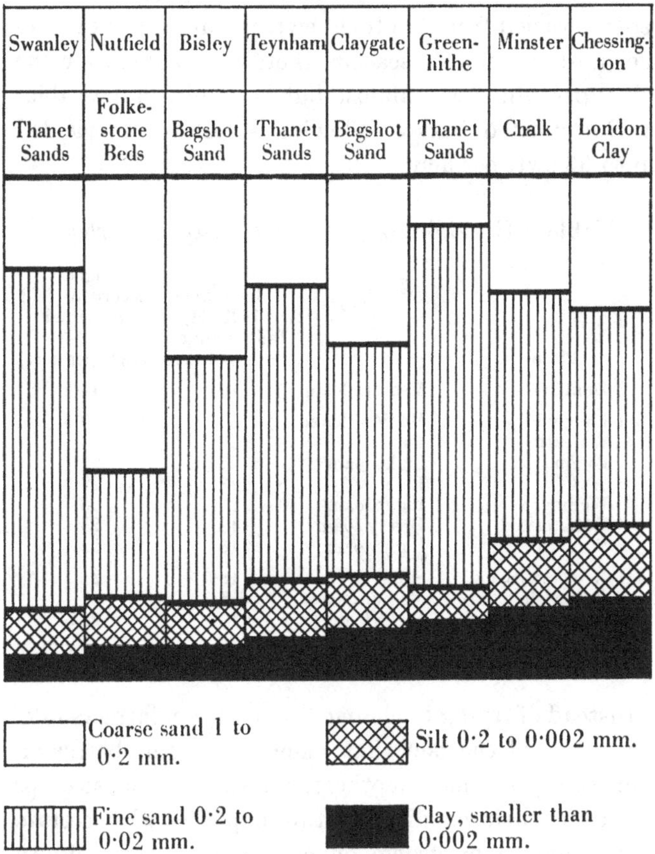

Swanley	Nutfield	Bisley	Teynham	Claygate	Green-hithe	Minster	Chessing-ton
Thanet Sands	Folke-stone Beds	Bagshot Sand	Thanet Sands	Bagshot Sand	Thanet Sands	Chalk	London Clay

Coarse sand 1 to 0·2 mm.

Silt 0·2 to 0·002 mm.

Fine sand 0·2 to 0·02 mm.

Clay, smaller than 0·002 mm.

Fig. 34. Mechanical composition of soils well adapted for potatoes.

factors that cannot be altered, though their effects may be mitigated: the soil type and climate. The effect of soil type is shown in Table XII, giving the results of a series of uniform experiments on barley carried out from Rothamsted under the Institute of Brewing scheme. The mean yields on the fully manured plot at the different centres varied from 9·5 to 40 cwt. per acre, and even in the most favourable seasons there was a limit for the soil types which was almost but not quite impassable: at Rothamsted barley following a fallow has yielded up to 35 cwt. per acre.

Table XII. *Effect of soil type on yield of barley*

	Place	Soil	Mean yield 1922–31, fully manured (cwt. per acre)	Highest recorded yield in period 1922–31 (cwt. per acre)
I	Dunbar	Medium loam	40	44·5
II	Longniddry	Medium loam	30	30·5
	Walcott	Fen	27	30
	Eyton	Light loam	25	27
	Wye	Loam	23	27
	Wellingore	Light loam	23	30
	Rothamsted	Heavy loam	19	25
	Woburn	Light loam	18	29
III	Newport (Salop)	Sandy loam	18·5	21
	Cawkwell	Medium loam	18·5	21
	Chiselborough	Light loam	17	21·5
IV	Martlesham	Light sand	9·5	12·5

Instead of trying to change the soil type, farmers take advantage of the fact that some crops grow better on some soil types than on others, and so they choose crops suited to their conditions; this explains why certain crops tend to be grown on certain types of soil. In Figs. 33 and 34 are shown mechanical analyses of the soils on which in the south-east of England wheat and potatoes are found to do well.

CLAY SOILS

Geologists distinguish a number of different clays: the London Clay, Wealden Clay, Oxford Clay, Lias Clay, etc., and these differ in their agricultural properties. They have, however, certain properties in common: they are sticky when wet but set very hard when dry, they swell up on moistening and give out a little heat, they absorb heat and shrink on drying, and thus form the large gaping cracks seen in dry weather on clay land. The fine particles also impede the movement of water so that the soil is very wet in wet weather but may suffer from drought in very dry weather.

These conditions are unsuitable to many plants but not to all, and the vegetation of the clay soils is restricted, but the plants that grow there do very well. The native flora consists of trees and bushes, particularly oak, ash, thorn, and in some places hornbeam[1]; clay land left to itself soon reverts to forest. At Rothamsted part of a wheat field was left unharvested in 1882: it was fenced off and has been left untouched ever since. Within thirty years it was covered with a dense growth of self-sown trees.

The lighter clay soils were early settled by farmers though the heavy ones were beyond their power of management and so remained in forest. Wheat and beans were found to do well and became the usual arable crops: the difficulty about drainage was overcome by laying up the land in great ridges and deep furrows (p. 117).

In the economic depressions of the last century many

[1] E.g. in Hertfordshire.

of the clay soils went out of cultivation and "tumbled down" to rough pastures many of which can still be seen. On these the surface-rooting bent grass (*Agrostis stolonifera*) is common: it withers during dry weather and causes the burnt colour so characteristic of poor clay pastures; rushes, the coarse file-like *Aira caespitosa*, and other plants specially adapted to wet places (Fig. 35) also flourish.

Fig. 35. Poor clay country. Roads wide but not all made up, hedges and gates not well kept.

There is, however, no justification for leaving clay lands in this condition. The drains and ditches should be cleared, the land ploughed, limed and manured and good crops can then be grown; plants suited to the conditions grow well because of the good moisture supply usually available. Ripening, however, is apt to be delayed, and the harvesting of cereals may be late.

Winter wheat and beans are still among the best crops; kale and mangolds also do well: potatoes, sugar beet, and swedes are not so good: barley grows well but is more often of feeding than of malting value. A summer fallow is the best treatment for the soil; a long autumn fallow comes next. However, even with modern implements, cultivation is not easy. For these reasons clay soils are generally laid down to grass. Existing pastures, even when poor, can often be much improved by basic slag, but it may be more economical to plough them out, add lime and basic slag, and sow with a mixture including good strains of rye grass, cocksfoot and wild white clover. The success of the scheme turns on the proper development of the clover, and a good supply of phosphate is essential to this.

Clays containing much silt are very difficult to deal with and no reliable method has yet been evolved. They can be found in the Lower Wealden Beds in the district east of Horsham, on the Boulder Clay, the Coal Measures, etc.: everywhere they have a bad reputation which they thoroughly deserve. Lime and subsoiling have less effect than might be expected, and probably the best treatment is to mole drain them and lay them down to grass: it is not worth while spending much on them as they do not respond well to treatment.

SANDS

The chief agricultural properties of sandy soils arise from the fact that they are porous and readily allow the passage of water. Thus the water never accumulates and the soils only get waterlogged when they are underlain by a basin of clay; usually they suffer from drought in dry weather. In its passage the water carries with it much of the soluble matter, sometimes indeed so much that even weeds will not grow but only patches of moss which decay to a black acid substance entirely unsuited to most plants; such patches can be seen frequently on the Bagshot Sands in Surrey.

Where there is a fair admixture of silt the movement of the water is retarded, and on moving aside the top 2 or 3 in. of soil the lower part is found to be quite moist even in dry weather. In these cases plants will grow well and special treatments have been evolved to suit them.

In the first place the movement of the water has to be still further retarded, and regular additions of organic matter are therefore necessary. Secondly, lime has to be added regularly except in certain special cases, e.g. where the soil lies at the foot of a long gradual slope and receives an underground drift of hard water from above. Lastly, fertilisers have to be added in small but frequent doses when the crop needs them. When these precautions are taken sandy soils will grow almost any crop, but they especially favour the development of roots and tubers so that they are well adapted to potatoes, sugar beet, turnips, carrots, parsnips and nursery stock; further, in suitable seasons they give good quality barley and useful but not large wheat crops. They are not suited for grass unless the water table happens to be from 3 to 4 ft. of the surface in which case they may carry magnificent pasture: some of the very best Romney Marsh pastures are on sand. Otherwise the grass burns up badly in the summer owing to lack of water.

Some sandy soils tend to form pans, and care has to be taken to prevent this by occasional use of the subsoiler.

The management of sandy soils turns on the method by which the organic matter is to be added.

(1) If stable manure is available in sufficient quantities a succession of heavy crops can be obtained, and recourse is then often had to market gardening: this is done on the sands near London, around Sandy in Bedfordshire, in parts of Essex and elsewhere. Where the market facilities are not quite so good, potatoes can be grown on a dressing of 12–15 tons of stable manure, and a mixture of artificials rich in potash; a grain crop, a "seeds" crop, and another grain crop can then be grown on the residues and without much further manure beyond nitrogenous top dressings.

(2) The organic manure may be supplied through the agency of livestock. In the old days sheep were kept on the arable land throughout the winter and folded on to green crops such as rape, kale, winter barley, swedes, vetches, etc.; in addition they received purchased feeding stuffs. The droppings from the animals fertilised the soil and returned to it a considerable part of the substance of the crops and feeding stuffs supplied. Moreover, the trampling of the animals had the further advantage of consolidating the land. This method was so good that farmers spoke of the "golden hoof"; unfortunately, it is now too costly for general use, though sugar-beet tops can still be fed on the land.

Another source of organic matter is the farmyard manure obtained from bullocks fed on home-grown fodder supplemented by purchased feeding stuffs; the manure, which contains much of the straw grown on the farm, is carted out on to the land.

(3) A third method of adding organic matter to the

soil consists in ploughing in a leafy crop: this is known as green manuring. It has answered well on very light sands: lupins fertilised with potash, phosphates and lime have been grown and then ploughed in; being leguminous plants they fix nitrogen from the air and thus increase the stock of nitrogenous organic matter in the soil, like a dressing of farmyard manure. An alternative method is to grow mustard, dress it with cyanamide and plough it in.

Green manuring, however, has not been extensively practised in this country because farmers prefer to feed their crops to stock and so get fat animals as well as manure. Catch crops are sometimes grown to be fed off or ploughed in before the regular crop: this method was developed to a high pitch of perfection on the sheep farms of the southern and eastern counties; elsewhere, however, it was not particularly useful.

Whatever the system of agriculture, it is desirable to crop as frequently as possible because sandy soils lose a great deal of their fertilising constituents if left bare and exposed to the rain. Weeds also must be kept down; no soils are so prone to be smothered with weeds as are sands.

The manuring has to be decided by the crop: organic matter and lime are of paramount importance, potash is wanted for many crops, especially potatoes, while phosphates are usually needed on the glacial sands of the eastern counties.

Sandy soils and light soils generally are attractive because they are more under control than most others. No matter how wet the season they can be worked. In many places it is possible to obtain two crops in the year: early potatoes may be followed by cabbage, sprouts, or sprouting broccoli. Strawberries can be successfully grown and many other valuable crops. Few soils, however, are so dependent on the skill and intelligence of the farmer. Some of the best farms in England are to be found on the sands: they are well managed, well manured, kept free from weeds, and made to yield heavy crops; labour-saving devices are introduced and the skilled hands are well paid. On the other hand, bad management speedily ruins the land and the farmer: docks, bindweed, sorrel, corn marigold, spurrey, ragwort and a host of other weeds soon come in and before long the land is useless.

LOAMS

Loams come in between sands and clays and are best defined as soils which are not as heavy as clays and not as light as sands. Usually they contain not more than 10–15 per cent of clay and not more than 20 per cent of coarse sand; they are chiefly made up of intermediate material. All shades of loams exist, from the light loams which some would call sands, to the heavy loams which can also be called clays.

Table XIII summarises the usual conditions on the heavy loam at Rothamsted; this is given for reference only.

Loams are by far the most fertile soils in the country; instances are to be found in the brick earths of east Kent and near Chichester, the alluvials of some of the famous vales and of the Evesham district, the famous Carse of Gowrie (locally called a clay) and many others. Practically any crops will grow—climate permitting,

Table XIII. *Conditions normally obtaining in the soil at Rothamsted**

Manurial treatment of soil	Cultivation treatment	Soil† moisture per cent by		Total nitrogen		Nitrogen as nitrate		Numbers of bacteria millions per gram	CO_2 per cent by volume in soil air
		Volume	Weight	Parts per million	Lb. per acre	Parts per million	Lb. per acre (top 9 in.)	Total (direct count)	
No manure	Cropped	23–12	15–8	990	2500	5–12	12–30	1000	0·2–0·6‡
	Fallow	23	15			8–15	20–36		0·2–0·4
Farmyard manure	Cropped	30–15	17–8	2200	5000	10–20	25–50	3000	0·5–1·0§
	Fallow	30	17			20–5	50–84		0·2–0·6
Ordinary arable field	Cropped	25–12	15–8	1500	3700	10–15	25–36	2000	0·5–1·0‖

* For pore space see p. 38.

† The concentration of the dissolved matter is of the order of 0·2 per cent, and the osmotic pressure about 1 atmosphere.

‡ Running on occasions up to 1·8.

§ Occasionally up to 2·5.

‖ Occasionally up to 2·3.

of course—and a wide range of farming systems is possible.

In the older systems bullocks or dairy cows played the central part on the heavy loams and sheep on the light loams, the animals in both cases being required to act as manure-making machines, and also to convert the less portable products such as straw, roots, etc., into portable and saleable meat. On the lighter loams the traditional rotation was clover, wheat, swedes, barley: the swedes and the aftermath of clover were fed off by sheep which also received cake and corn; the wheat and barley were sold. On heavy loams the rotation was somewhat different: clover or beans; wheat, mangolds (with some swedes), oats (and some barley)—swedes and barley, being less suited than mangolds and oats for heavy land, were less grown. Dairy cattle and bullocks were kept. Nowadays, however, many farmers have no fixed rotation but grow those crops that promise to be profitable at the time. Among the crops grown in the eastern counties are sugar beet, potatoes, Brussels sprouts, kale, mangolds, peas and lucerne. The best quality malting barley usually comes from the loams. The keeping of pigs and poultry has greatly expanded.

The lighter loams are very prone to weeds, in particular to charlock (*Brassica sinapis*). Heavy loams suffer less, but still are liable to docks, thistles, etc. Charlock can be kept down by spraying,[1] the others cannot. Sometimes the land will keep clean for four and sometimes for five years: in that case two corn crops can be taken in succession and a winter oat crop inserted

[1] A 3 per cent solution of copper sulphate sprayed in early spring at the rate of 50 gallons per acre. Other chemicals are available also.

between the wheat and the roots: or the clover may be replaced by a mixture of clovers and grasses which can be left for a period of years.

The root crop affords the best means, next to fallowing, of cleaning the land. It is usually taken after a corn crop, so that the land can be well cultivated from November or December to the time of sowing; cultivation can continue till June, when the root crop begins to grow; or if necessary even longer. The grain crops, on the other hand, follow continuously: the barley is seeded with clover so that the land is not even ploughed between these two crops: the clover is ploughed in just before the wheat is sown, and if winter oats follows, this crop in turn is sown just after the wheat is harvested. Only when the root crop comes round is there much opportunity for cultivating the soil well and giving it the benefits of the partial fallow.

In forward districts the harvest may come so early that the tractors can at once be put on to the land and a bastard fallow given before the next corn crop: it is then not necessary to grow roots but a series of corn crops can be taken, with occasional clover crops.

This system can be worked with a minimum of live stock and of horses, also of men. It is the basis of mechanised farming. Its disadvantage is that fungi causing plant diseases are apt to get into the soil and to stop there, being favoured by the almost continuous presence of their host plants. Among the more serious of these are the "take-all" disease of wheat (*Ophiobolus graminis*) (Fig. 36) common on chalky soils, and a lodging disease of cereals (*Cercosporella herpotrichoides*) (Fig. 37).

Fig. 36. A large Take-All patch at Karoonda, South Australia.

Fig. 37. Lodging of wheat due to the fungus *Cercosporella herpotrichoides*, Pastures Field, Rothamsted, 1937.

The clover or seeds mixture adds nitrogenous organic matter to the soil (p. 50), and so increases its fertility.

It is an old practice in the north of England and Scotland to leave the seeds mixture for three, four or more years, and this is spreading in other parts of the country.

Unfortunately, red clover cannot be grown by itself very frequently on the same land in England as it suffers from diseases and pests called generally "clover-sickness". The trouble may be due to lack of lime or of potash, but it may also arise from physiological or pathological causes. The soil should be tested for acidity: if this is not the cause a dressing of sulphate of potash (2 cwt. per acre) may be tried, and if this still fails another leguminous crop ought to be grown instead of red clover.

The lighter loams are much used for potatoes, sugar beet, and special crops like fruit, market garden and nursery produce, malting barley, etc. Some have always been used for these purposes, such as the Thanet Beds of east Kent, but many of them, like the sands, were formerly held in but little repute, and have only during the past 30 or 40 years come into favour. The light loams of the New Red Sandstone of Somerset are well adapted to fruit, market gardening, etc., while the light loams round Porlock are famous as the source from which many prize samples of barley have come to the Brewers' Exhibition.

CHALK SOILS

Chalk soils are usually very light loams, and like them they are liable to drought, but they possess the unfortunate property of drying to hard steely fragments unless they are worked to a good tilth at the proper time: they therefore require special care in cultivation. Organic matter is very necessary for them, and in the old days sheep therefore played a large part in chalk districts both for supplying manure and for treading the soil. Further, during frosty weather they become so puffed up and lightened that the young crops are sometimes almost forced out of the ground; rolling is therefore necessary in the spring not only on the grass but also on the arable land.

Leguminous crops are especially valuable on the chalk by reason of the organic matter they introduce; among the most useful are sainfoin and lucerne, the latter especially in the drier regions where the subsoil is well drained.

Chalk soils are highly favourable to plant and animal life, but this has disadvantages: they carry a very varied flora and care is needed to keep down weeds, especially charlock. Swedes and other brassicas are liable to attack by the turnip fly (*Phyllotreta nemorum*), and all crops to damage by wireworm.

The central feature of the manuring used to be the folding of sheep: superphosphate was given to the roots, and potash manures for the clover or seeds ley. No better system is known, but unfortunately folding is no longer profitable and chalk farmers are often in difficulties about the maintenance of fertility. On the large fields of the open downs a good deal of mechanised farming is practised, but the conditions favour "take-all".

Some interesting farming developments have taken place on the chalk grassland, formerly regarded as sweet but poor. Somerville showed that basic slag often effects remarkable improvements, especially in the wet districts or where the top inch or so of soil has lost its calcium carbonate, while Hosier has worked out a system of management of dairy cattle and poultry that greatly enriches and improves the herbage.

FEN AND PEAT SOILS

Three kinds of these soils occur in Great Britain:

1. The fen soils of Huntingdonshire, Cambridgeshire and Norfolk, forming a large area of low-lying land that would be flooded by the rivers but for the elaborate system of embankments and pumps. The rivers and seepage waters bring in much calcium carbonate and hence the organic matter is rich in bases, it is much less acid than the peats, and it is rich in nitrogen. Hence the soils are very fertile. They produce heavy crops of potatoes, sugar beet, oats and wheat; also mustard for seed, celery and buckwheat. The fen that lies over Kimmeridge Clay (Isle of Ely) is greatly benefited by dressings of the clay brought up from below the surface. The method consists in laying out trenches, 18 yards apart, and in each digging holes ten to the chain and sufficiently deep to reach the clay: about a ton is then obtained from every hole and spread round about. The cost of the operation at the time when much of it was done (i.e. prior to 1914) was about 50*s*. per acre.

The fen remote from the clay is not so readily improved: corn crops, for example, yield less grain. In cultivating fen soils the great need is to keep down the weeds and leave the land sufficiently compact. Oxidation and erosion are rapidly taking place, and even within living memory have caused much shrinkage of the fen.

Generally speaking lime is not required for crop production, probably because the waters feeding the fen streams come from calcareous regions, and therefore bring with them large amounts of dissolved calcium carbonate. The chief fertiliser needed is superphosphate, of which large dressings are sometimes given with profitable results. Potatoes, however, respond to nitrogen and to potash.

Little livestock is kept and but little farmyard manure is therefore available: it gives, however, good results on the lighter soils.

2. *Low-lying peat lands.* These occur to a considerable extent in the western half of England, Wales and in Ireland. They receive from incoming water much less calcium and other basic material than the fens, and so are more acid; indeed they need lime. Two general methods of treatment have been adopted: the peat is dug out and sold as fuel, then the underlying ground is cultivated, or the peat is ploughed and cultivated direct. Drainage is a first essential. Oats, potatoes, buckwheat and many grasses will grow well, but lime is needed for almost any crop, and in many cases potash as well.

The cultivation of this type of land has been reduced to a fine art in Holland, Belgium, and north-west Germany: companies have been formed for the purpose of reclaiming areas previously waste. The general method of procedure is to drain, then plough deeply, to add sand or lime, leave for a time to the action of the weather, then plough again, add lime and the proper artificials, and finally harrow down, when a good seed bed can be obtained. In some cases, however, crops fail for lack of copper and a small dressing of copper sulphate—about 20 lb. per acre—must be added. It is almost solely on these soils that copper deficiency is observed (p. 201).

3. *High-lying peat land.* In the northern counties there are considerable areas of moorland at high altitudes which, however, seem wholly unsuited to cultivation. The rainfall is high and evaporation is low; the calcium and other bases are washed out, leaving a very acid residue quite unsuitable to crop production until large quantities of lime are added. Pastures have been obtained by enthusiastic reclaimers at considerable cost, but they are kept only with difficulty; the winters are inclement and drainage is a serious trouble. Experiments are being made on this kind of peat in the Island of Lewes by the Macaulay Institute, Aberdeen.

UPLAND SOILS

In high-lying parts of the north and west of Great Britain the high rainfall and rather cool temperature govern the situation and dictate a predominantly grass agriculture with only the minimum of arable cropping. As would be expected (p. 56), the soils are acid, hence the arable crops include oats, potatoes, rye if wanted, but only selected varieties in any case should be grown. Unless the soil is covered by a crop it loses plant food by leaching and grass is the best preventive against this. Cattle- and sheep-rearing are commonly practised, and milk production where transport is possible; butter-making used to be more common than it is now (Fig. 38).

Improvement of the herbage commonly requires drainage and application of lime and slag, and many cases are on record where nothing more has been

Fig. 38. The contrast between upland and valley land; the highest land is much leached, the soil is acid and infertile. The slopes receive some plant food and carry vegetation. The lower land is the most productive, provided it is drained and limed. (Mynach Valley, Cardiganshire.)

needed. But Stapledon has shown on the Cahn Hill Farm that much quicker and better results are often obtained by ploughing up, then adding lime and slag, and sowing with a mixture of selected grass and clover seeds (Fig. 39).

SUMMARY

In order to get the best out of the land an examination must be made to see what are likely to be the chief defects, in other words, what will constitute the limiting

Fig. 39. The Cahn Hill experimental farm, Cardiganshire. The dark-coloured areas represent the native vegetation. The light-coloured areas have been broken up, manured and sown with selected grass and clover seeds. Animals prefer these areas and remain on them.

factors. There may be insufficient water or excess of water, inadequate depth of soil, lack of any of the proper constituents: (*a*) of clay, when the soil will not hold together but will blow about; (*b*) of calcium carbonate, when the soil will be acid as shown by the presence of sorrel, the failure of clover, and by poor growth generally; (*c*) of organic matter, when the tilth will be unsatisfactory; (*d*) of various nutrient substances.

The defect may arise from the fault of the soil itself or of its situation.

Any defect of this kind will set a limit beyond which the crop cannot be increased. Until this is removed it is useless to try and force the crop beyond the limit thus set. Once it is removed, however, better crops can be obtained.

The central features of management in cropping land up to its full capacity are:

Crops and varieties are selected that are specially suited to the conditions. Some crops, such as buckwheat, rye and flax, will tolerate poor conditions; others will not.

Sufficient lime is added and the land is properly cultivated. Every effort is made to keep up the supply of nitrogenous organic matter in the soil: leguminous crops are grown; "seeds" are left for two or three years and sometimes crops are grown simply to be ploughed in. The supply of plant nutrients is kept up by the addition of appropriate fertilisers and by supplying imported foods to sheep, pigs and cattle, when much of the fertilising constituents are excreted and can be put on to the land.

Part of the land is kept in permanent pasture and thus becomes richer in nitrogenous organic matter. The necessary mineral food is added in the form of calcium phosphate and potassium salts.

PART III

FERTILISERS

CHAPTER VII

THE FOODS OF PLANTS

It has taken some 350 years of experiment to find out what are the foods of plants, and even yet our knowledge is very incomplete; something new is continually being discovered.

The subject has been approached in several ways. The early investigators started out with the idea that a certain substance was the food of plants, or, as they called it, "principle of vegetation", and then proceeded to see if this was so. The first really good experiment was made in Brussels early in the seventeenth century by Van Helmont (1577–1644), who thought that water was this "principle". His account constitutes one of the shortest but most important scientific papers ever published: "I took an earthen vessel in which I put 200 pounds of soil dried in an oven, then I moistened with rain water and pressed hard into it a shoot of willow weighing 5 pounds. After exactly five years the tree that had grown up weighed 169 pounds and about three ounces. But the vessel had never received anything but rain water or distilled water to moisten the soil when this was necessary, and it remained full of soil, which was still tightly packed, and, lest any dust from outside should get into the soil, it was covered with a sheet of iron coated with tin but perforated with many holes. I did not take the weight of the leaves that fell in the autumn. In the end I dried the soil once more and got the same 200 pounds that I started with, less about two ounces. Therefore the 164 pounds of wood, bark, and root arose from the water alone."

The facts are correct, and the conclusion seems to be absolutely sound; it satisfied the best scientists of the time. Nevertheless it is quite wrong, and the experiment stands as a lasting reminder that we see things only "through a glass, darkly", and can never be sure that our explanations are correct.

Somewhat later Glauber, a German, believed that saltpetre (nitrate) was the "principle". He used some of it as fertiliser with very good results; he showed, moreover, that it could be extracted from the earth cleared out from cattle sheds, and argued that it must have come from the animals' excreta and therefore from their food. This was confirmed by an English chemist, Mayow, who showed by analysis that natural soils contain saltpetre in spring when plants are just beginning to grow but not when growth is vigorous; "the reason being", he said, "that all the nitre of the soil is sucked out by the plants".

An entirely new method was adopted by another Englishman, John Woodward, at the end of the seventeenth century. Starting with Van Helmont's experiment, and apparently not knowing about the work on saltpetre, he grew a plant, spearmint, in water and showed that something more than water was concerned: River Thames water was a better food than rain water, while Hyde Park conduit water (which many Londoners drank at that time) was better still, and water that had been shaken up with garden soil was best of all. This, he said, showed that "terrestrial matter" and not water was the food of plants. This method has been much developed and is now widely used in botanical laboratories under the name "water culture".

Some sixty years later, in 1757, a Scotsman, Francis Home, made the first pot experiments. Their advantage was that he could easily test the effect of various substances on plant growth. His actual results were not particularly good, but the method, like Woodward's, has remained and is now adopted in nearly all fertiliser investigations.

Meanwhile chemistry was developing, and it was shown that men and animals breathe out a gas called carbon dioxide which is poisonous to themselves. Priestley, a Birmingham minister, realised that there must be some purification process in Nature or life would long since have become impossible. He kept a mouse in a closed jar till it became affected by the vitiated air, then put in a sprig of mint and found that the air became purified. Plants must therefore be the restoring agents, and it is through them that animal life continues on this earth. Later on it was shown that plants also breathe out the same gas as animals and so vitiate the air; this caused much mystification till the Dutch physician Jan Ingenhousz, working first in Vienna and then in London, and the Swiss botanist Senebier at Geneva, showed that purification of air by plants goes on in daylight only, while vitiation by plants goes on in darkness—and as is now known, in daylight as well.

Little more could be done, however, till chemistry was more advanced. Before long the stage was reached when another Geneva botanist, de Saussure, in 1804 was able to straighten out the whole tangle. The vitiation of air was by that time known to be due to animals breathing in oxygen and breathing out carbon dioxide which is poisonous to them: de Saussure showed that plants in daylight, but not in darkness, took in this carbon dioxide, assimilated the carbon and built it up into their tissues, liberating the oxygen and thus restoring the air to its original purity. At the same time plants, like animals, were also breathing in oxygen and breathing out carbon dioxide, but their purifying effect was greater than the vitiation they brought about.

A new method of approach was meanwhile being developed. Chemists were discovering how to analyse plants and soils so as to find out the substances present in them. It was assumed that if a substance occurs in the plant it must be serving some useful purpose and therefore must be supplied in the plant food. Later work showed that this is not quite correct, nevertheless the idea served a useful purpose. The new chemical methods were used to great advantage by the French agriculturist Boussingault, who was the first to set up a laboratory on the farm and to make really scientific field experiments. He can be regarded as the founder of agricultural science. On his farm at Bechelbronn in Alsace he carried out a magnificent series of experiments in all branches of agriculture. In one of these he weighed and analysed all five crops of a rotation,[1] so as to discover what the plants had taken up: he analysed also

[1] Beets; wheat; clover; wheat; turnips (catch crop); oats.

the farmyard manure so as to see what it had supplied, and so by difference discovered the amount of substances that had come from the soil and the air. The results were, in cwt. per acre:

	Dry matter	Carbon	Oxygen	Nitrogen	Mineral matter
In the six crops	140	65·6	56	2·03	8·5
In the manure	81	29·1	21	1·63	26·2
Difference taken from soil and air	59	36·5	35	0·4	- 17·7

This shows that the crop consists very largely of carbon and oxygen, of which the manure could have supplied only a part, and as the soil was in approximately the same state at the end of the rotation as at the beginning it was assumed that the rest had not come from there. Later work shows that this is so. Both carbon and oxygen come from the air, and as these form some 90 per cent of the dry weight of the plant we now have the true explanation of Van Helmont's result: it was the air and not the water that gave the increase. In his day nothing was known about the air, and its part in plant growth was not suspected.

The crop took out rather more nitrogen than had been supplied, so that there was a gain on the whole rotation. This could not be explained for many years, but then it was found that the leguminous crop in the rotation fixes a certain quantity of gaseous nitrogen from the air and leaves some of it in the soil for the next crop.

The mineral matter taken by the crop is only small in amount and less than in the manure.

Chemical analysis has also been used in another way. Farmers found by experience that certain substances were of great value as manure. Chemists analysed these, and found that they contained some of the substances present in the plant, particularly nitrogen, phosphorus and potassium. Liebig, the German chemist, in 1840 brought all these scattered results together and showed that the mineral matter, small though it was in amount, was indispensable to the growth of plants and must be supplied in sufficient quantity: it could be given in the form of simple salts.[1] Finally, Lawes at Rothamsted showed how this could be done in farm practice. He set up the first factory for making plant foods, or, as they became called, artificial fertilisers. In association with Gilbert he started the field experiments which are still continued and which have given us much of our knowledge of the properties of fertilisers and the best way of using them.

The food substances added to the soil fall roughly into two groups: fertilisers and manures. The distinction between the two terms is not very sharp, but generally a fertiliser is a concentrated substance imported on to the farm from a foreign country or a factory, and therefore is frequently called an artificial fertiliser, while a manure is a more bulky material either produced on the farm or closely related to farm products.

The substances thus added to the soil are compounds of nitrogen, phosphorus and potassium: also organic matter and lime or chalk. In order to study their effect on the soil a series of pot experiments should be started:

[1] Note the contrast between plants and animals. The plant lives on simple substances, carbon dioxide and salts, building them up into complex "organic matter" containing energy derived from the sun's light. Animals, including ourselves, live on these complex substances and derive all their energy from them. Plants build up, while animals break down.

10 in. flower pots are sufficiently good for ordinary purposes, but for finer work Doulton's glazed pots must be used (Fig. 1). The soil has to be carefully mixed to ensure uniformity, and if it is heavy 10–20 per cent of sand must be added. The series should contain pots treated as follows: (1) unmanured; (2) and (3) 0·01 and 0·05 per cent respectively of nitrate of soda; (4) 0·05 per cent superphosphate; (5) 0·025 per cent sulphate of potash; and three or four containing combinations of these quantities; other pots should be supplied with sulphate of ammonia or nitro-chalk in place of nitrate of soda, and bone meal and basic slag in place of superphosphate. If a glasshouse is available tomatoes are a good crop for experiment; or, at colder seasons, mustard. For outdoor work rye, wheat or mustard do well.

Where land is available eight small plots can be laid out and given the following dressings[1]:

(1) No manure; (2) nitrogenous manure (N); (3) phosphate (P); (4) potassic manure (K); (5) NP; (6) NK; (7) PK; (8) NPK.

The nitrogenous manure (N) may be nitrate of soda given at the rate of 2 cwt. per acre = 2 oz. per 3 sq. yd.; the phosphate (P) may be superphosphate at 6 cwt. per acre; and the potassic manure (K) may be sulphate of potash at 3 cwt. per acre. Lettuce makes a good crop.

For many years it was believed that these elements represent all that need be added to the soil in order to ensure full plant growth. Then quite unexpectedly it was discovered that a whole series of other elements are needed but only in minute amounts: they had been missed because the older compounds of nitrogen, phos-

[1] For the proper performance of this experiment see p. 272.

phorus and potassium had been impure and had contained them unknown to the experimenters. When pure substances were used they failed to satisfy the plants' requirements, and certain "deficiency" diseases developed.

This is the newest branch of the subject and one of the most interesting: it is dealt with in Chapter XI. Even now there is still much to learn about plant food, and investigations are continually being made at the Research Institutes where these problems are studied.

CHAPTER VIII

THE NITROGENOUS FERTILISERS

THE two most striking effects of nitrogenous fertilisers on crops are the deepening of the green colour of the leaves and an increased rate of growth. A nitrogen-starved plant is yellowish and stunted; when the fertiliser is added the change in colour can be seen almost in 24 hours in favourable conditions. The increased rate of growth is caused by an increase in leaf size, but not in leaf efficiency per unit area. Equal-sized leaves of unmanured plants and of plants manured with a moderate quantity of one of these fertilisers are equally efficient in fixing carbon dioxide and converting it into sugar, but the manured plants, having more and larger leaves, can produce more sugar in a given time.

The amount of plant material that can be built up by an additional supply of nitrogen to the crop is considerable. Lawes and Gilbert showed that 1 lb. of nitrogen supplied as manure adds between 40 and 50 lb. of dry matter to the crop of cereals, mangolds and sugar beets, though less for the swedes and potatoes of their day. The increases are more consistent, and vary less from soil to soil and from season to season, than those given by phosphates, potassic salts or any other fertiliser; indeed, of all fertilisers the nitrogenous are the most certain in their effect. Some of the increases obtained are given in Table XIV.

Additional fertiliser often gives additional crop, but beyond a certain point the yield *per cwt. of manure* falls

Table XIV. *Increases in amounts of crop obtained by applying* 1 *cwt. sulphate of ammonia* (23·5 *lb. nitrogen*) *per acre in presence of sufficient phosphate and potash*

	Cwt.		Cwt.
Wheat grain	2·5	Swedes	17*
Barley grain	3	Swedes	47†
Oats grain	3	Sugar-beet roots	10–15
Hay	4–6‡	Sugar-beet tops	15
Mangolds	30	Kale	30
Potatoes	20		

* South of England (Rothamsted).
† North of England (Cockle Park, 1902–4, and Durham Exp. 1904).
‡ 4 cwt. at Rothamsted with sulphate of ammonia. G. Turnbull (*J. Min. Agric.* 1920, **26**, 611) estimates 5 or 6 cwt. for nitrate of soda. He states that sulphate of ammonia has sometimes given considerably more.

off, so that the extra crop is obtained at higher cost per ton or per bushel than a smaller crop would be. This is known as the Law of Diminishing Returns, and it holds very generally; it is just as true for the horse-power of a motor cycle as of the yield of wheat. Fig. 40 gives an illustration from the Broadbalk plots at Rothamsted.

The yield of straw continues to increase almost at the same rate throughout, but the increased yield of grain is much less for the third dose of nitrogen than for the first or second. This means that the leaves of the heavily manured plants are less efficient than those receiving a moderate dressing, for although the third dose of nitrogen gives the same weight of straw as the second, it does not produce as much grain.

The appearance of the leaf is also changed: it becomes much darker in colour, and in some cases crinkled. Not infrequently the head and leaf of a cereal crop becomes so heavy that the straw is unable to carry the weight in wet weather and the crop becomes laid. The plant also becomes more liable to the attack of certain fungi: one of these not infrequently attacks the base of the

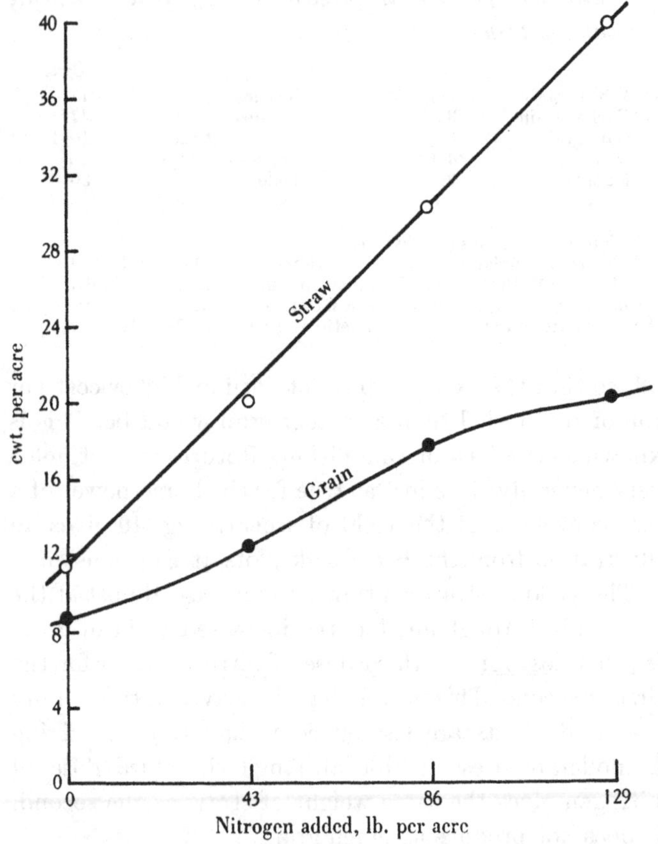

Fig. 40. Yield of wheat, grain and straw. Broadbalk field, Rothamsted.

stem of wheat, causing it to lodge; others attack the leaves of mangolds receiving much nitrogen and insufficient potash.

The increases are expressed in terms of feeding-stuff units in Table XV, and this shows that about 3 lb. of protein and 10–40 lb. of total dry matter are obtained per lb. of nitrogen applied in the manure. Expressed in other terms 1 cwt. of sulphate of ammonia may be expected to give a return of about 2 stones of protein equivalent, and about 20 stones of starch equivalent.

Table XV. *Food value produced by a nitrogenous fertiliser (sulphate of ammonia)*

	Usual increase per 1 cwt. sulphate of ammonia	Protein equivalent per cent	Starch equivalent per cent	Produced by 1 cwt. sulphate of ammonia: Protein equivalent lb.		Produced by 1 cwt. sulphate of ammonia: Starch equivalent lb.	
Potatoes	20 cwt.	0·6	18	13		403	
Mangolds	32 ,,	0·4	7	14		251	
Swedes	20 ,,	0·7	7	16		157	
Kale	30 ,,	1·3	9	44		302	
Barley:							
Grain	3¼ ,,	6·2	71	23	28	258	391
Straw	6¼ ,,	0·7	19	5		133	
Oats:							
Grain	2½ ,,	7·6	60	21	27	168	282
Straw	6 ,,	0·9	17	6		114	
Wheat:							
Grain	2½ ,,	9·6	72	27	33	212	274
Straw	5 ,,	0·1	11	6		62	
Meadow hay	5 ,,	4·6	31	26		174	
			Mean	25		279	

The values for the percentages of protein equivalent and starch equivalent are taken from Rationing of Dairy Cows, *Dept. Cttee. Rep. Min. Agric.* 1925. The money values can be calculated from the prices quoted in the *Journal of the Ministry of Agriculture.*

Barley and potatoes are most efficient as transformers of cheap fertilisers into food: kale and oats come next.

These figures, however, represent only the average

effects of fertilisers: there are many factors that modify their action in particular cases.

Rainfall is perhaps the most important of these. Nitrogenous fertilisers act best under a rainfall of 22–40 in. per annum. In drier conditions they are much less effective and may even do harm by causing too great a growth of leaf. In wetter conditions they are liable to cause cereals to lodge and do not seem to have their full effect on potatoes, though this is not certain.

Late-sown crops also benefit less from nitrogenous fertilisers than those sown at the proper time.

The effect of the nitrogenous fertiliser depends on the quantity given. Additions of 20 or 30 lb. of nitrogen per acre (e.g. 1–1½ cwt. sulphate of ammonia) increase the size of the plant without much change in composition. The additional nitrogen causes the leaf to become larger and so it assimilates more carbon dioxide and produces more carbohydrate, and this usually just about balances the nitrogen taken up. But when the quantity of nitrogen taken up exceeds a certain amount it is no longer fully balanced by carbohydrates and it then remains as an excess in the plant and may produce harmful effects. The leaves become dark green, crinkled, soft and sappy. Ripening is also delayed. The percentage of nitrogen in the plant increases: this lowers the malting value of barley, the percentage of sugar in sugar beet, the cooking quality of potatoes, and the wine-making quality of grapes.

Up to a certain point nitrogenous fertilisers increase the yield, especially of the leaves, almost in proportion to the amount given, but they do not always correspondingly increase the rest of the plant, e.g. the root of sugar beet and the amount of grain produced by cereals (p. 160). Their effect is most consistent on leaf crops, grass, kale, cabbage, etc., which they increase considerably.

Thus for sugar beet the results have been:

	Mean yield tons per acre	Increase in cwt. per acre for sulphate of ammonia	
		2 cwt. per acre	4 cwt. per acre
Tops	9·2	30	58
Roots	10·4	18	28
Sugar	1·8	3·0	4·1

The increase of tops for the double dressing is double that for the single dressing, but the increase in roots is only 55 per cent and in sugar is even less. Here the leaf has lost efficiency as the result of the nitrogenous manuring, as happened with the wheat in Fig. 40.

THE NITROGENOUS FERTILISERS IN COMMON USE

For many years the only nitrogenous fertilisers (apart from organic manures) in common use were nitrate of soda from Chile, and sulphate of ammonia, a by-product in the manufacture of coal gas. As the demand increased fears were expressed that the supplies would one day give out, leaving the world faced with starvation. This gruesome possibility was pointed out by Sir William Crookes in 1896 and, like a true man of science, he also showed the way out of the difficulty. The nitrogen of the air can be united with the oxygen of the air to form a nitrate, with the hydrogen of water to form ammonia, or with calcium carbide to form calcium cyanamide. All of these are effective fertilisers and the supplies are inexhaustible so that all fear of nitrogen starvation is banished.

Those in common use are:

	Formula	Nitrogen per cent: In pure salt	Nitrogen per cent: Usual in fertiliser
Nitrate of soda	$NaNO_3$	16·5	15·5–16
Nitrate of potash	KNO_3	13·85	15*
Nitrate of lime	$Ca(NO_3)_2$	17·8	15·5
Nitrochalk	$NH_4NO_3 + CaCO_3$	35†	15·5
Sulphate of ammonia	$(NH_4)_2SO_4$	21·2	20·6–21
Cyanamide	$CaCN_2$	35	20·6

* Some nitrate of soda is always present, which raises the percentage of nitrogen.

† In ammonium nitrate.

The nitrates

Nitrate of soda. Four nitrates are now available as fertilisers: the nitrates of soda, of potash, of lime and of ammonia, but of these the commonest is nitrate of soda. This substance occurs in the rainless regions of Tarapaca and Àntofagasta in the north of Chile, where it forms deposits near the surface of the soil, occurring in detached areas over a wide range. It is not known how the deposits originated: there is little doubt that they were once under water, but there is nothing to show how so much nitrate came to accumulate in one district: only traces occur in ordinary sea water. The crude nitrate is excavated by a process of trenching, it is then crushed, purified by recrystallisation and put up in bags for the market.

The commercial product is not quite pure, but it is guaranteed to contain 95 per cent of nitrate of soda and often contains more.

Among other substances present are boron, also an essential plant food, and iodine, a necessary food for animals.

Nitrate of soda is also prepared synthetically, and in this form it can be obtained in very pure condition. It then, however, is free from boron and iodine, and in well-conducted pot experiments can be used to demonstrate the necessity for these substances.

Nitrate of soda is very quick acting as a fertiliser and can be taken up immediately by the plant. It finds application in two cases: (1) in case of emergency, when young plants are suffering through an attack of a pest, or in cold wet weather; (2) in ordinary practice as a top dressing for the crop. It causes increases of practically all crops in England, and the dressing applied varies from 1 cwt. per acre to wheat in spring or grass laid in for hay, up to 10 cwt. per acre used on the valuable early cabbage and broccoli crop in Cornwall. It is, however, not used for main crop potatoes.

In drier conditions good returns are not always obtained; in parts of Australia and New Zealand phosphates are the limiting factor: in Western Canada water appears to be; in none of these cases do nitrates give the same returns as in this country.

The sodium is of value for mangolds and sugar beet, and these two crops are particularly responsive to nitrate of soda. It is very effective in increasing yields of hay: indeed, at one time farmers near to towns producing hay for sale frequently gave nothing else to the crop. This, however, was unwise: the other necessary plant foods should also be present. Nitrate of soda encourages a good growth of top grasses, probably by encouraging a deep-rooting habit, but the heavy crop naturally draws on the other plant foods in the soil, and unless these are replenished at the same time the soil becomes impoverished and the crop ultimately falls off in quantity, while weeds and poor grasses appear and bring down the quality. Grassland should never as a regular course be fertilised with nitrogenous fertilisers only, but should periodically receive the other necessary manures.

In addition to these effects on the plant certain effects are produced on the soil. The nitrate of soda is not taken up bodily by the plant but undergoes decomposition, the nitrate part being taken by the plant while some of the soda remains in the soil, and there conserves the reactive calcium, thus reducing the net loss of calcium carbonate and to this extent preventing the soil from becoming acid. It also causes more of the potassium compounds in the soil to become available for the plant; this has been demonstrated by actual field experiments at Rothamsted, and is also illustrated by the experiment on p. 173.

Nitrate of soda readily washes out of the soil and must therefore not be applied until it is needed. On the light soil at Woburn its superiority to sulphate of ammonia is more marked in dry than in wet seasons, though this is not so on the heavier soil at Rothamsted: indeed, the position there seems to be reversed. Usually it is best put on as a top dressing when the plant is up: when this course is adopted the loss in a wet season is reduced to a minimum. On the Continent it is often all applied with the seed for sugar beet. Heavy dressings such as are used in market gardens should be applied in two or three lots and not all at once.

Nitrate of soda does not mix well with superphosphate, although if put on at once the mixture can be used without harm. But as a rule the superphosphate should go on earlier.

Two forms of nitrate of soda are on the market, granular and crystalline. The difference lies in their mode of preparation; both are equally effective and contain the same percentage of nitrogen, 16 per cent, the guaranteed minimum being 15·5 per cent. The granular is the more usual form in Great Britain; some farmers find it easier to handle and to drill; in Egypt, on the other hand, the crystalline form is the common one.

Nitrate of potash. The pure salt is not used as fertiliser, being too expensive, but a mixture of about 1 part of potassium nitrate and 2 parts sodium nitrate is prepared in Chile under the name "potash nitrate". The quantities available are not large, and they mostly go to Hawaii, the United States,[1] and the South American countries; only a small amount comes to Great Britain. Here it is used chiefly for market-garden crops, and on light soils. It contains 15 per cent nitrogen (N) and potassium equivalent to 15 per cent K_2O; at the time of writing it is cheaper than a mixture of nitrate of soda and sulphate of potash containing the same ingredients. Where both potash and nitrogen are needed, therefore, it may prove very economical.

Nitrate of lime. Nitrate of lime is a manufactured product, the nitrogen and oxygen being obtained from the air and the lime from limestone. Great quantities of electric power are needed in the process, and the works are therefore situated in places where cheap water

[1] Chiefly for fruit and vegetables in the Atlantic and Gulf States.

power is available for generating electricity. The chief producing countries are Norway, Germany and France, and the consuming countries are Germany, Egypt, Scandinavia, France and Italy: only small quantities are used in England.

The old samples were apt to become sticky and to cake, making distribution difficult, but this trouble has now been overcome by mixing with a little ammonium nitrate, granulating, and storing in special moisture-resistant bags.

Nitrate of lime has no harmful effects on the soil and is useful to practically all crops. It contains about 15·5 per cent of nitrogen.

Ammonium nitrate: nitrochalk. Ammonium nitrate is one of the quickest acting and most economical nitrogenous fertilisers that can be made; it is also one of the most concentrated. When pure it contains 35 per cent of nitrogen, half of which is in the form of nitrate and half in the form of ammonia.

It has, however, two drawbacks: in its concentrated form it is deliquescent and liable to set in a hard mass: and there are conditions under which it may explode, but not, however, on a farm. In order to avoid both difficulties the ammonium nitrate is mixed with calcium carbonate in such proportion that the resulting product contains 15·5 per cent of nitrogen and 29 per cent of calcium oxide.[1]

The mixture is called nitrochalk, and as it is graded to contain the same quantity of nitrogen as nitrate of soda it is offered at about the same price.

[1] I.e. 44·3 per cent ammonium nitrate and 51·8 per cent calcium carbonate.

Ammonium nitrate tends to make the soil acid; it has only half the acidifying effect of an equal weight of sulphate of ammonia and only one third the effect of an amount containing the same weight of nitrogen. The calcium carbonate, however, completely counteracts this tendency, and nitrochalk can safely be used in all soils, including heavy and acid soils.

Sulphate of ammonia. Until about 1920 this substance was manufactured almost entirely from coal. One ton of coal contains on an average some 25 lb. of nitrogen, which if fully recovered would yield just over 1 cwt. of sulphate of ammonia. Unfortunately, most of the coal is burnt under such conditions that the nitrogen is lost, but in certain industries, especially in the manufacture of coal gas, of producer gas, in coking ovens, etc., special recovery methods are used and sulphate of ammonia is obtained as a by-product. The world's output from this source in 1937–8 was well over two million tons, this being nearly six times the quantity produced in 1900.

In recent years, however, increasing quantities of sulphate of ammonia have been manufactured synthetically. The amount thus made is now almost twice as great as that made from coal. The nitrogen is obtained from the air, the hydrogen from water, and the sulphate part either from gypsum or from by-product sulphuric acid, of which the smelting industries provide vast quantities. The process requires great technical skill and costly equipment but little electrical power: it is therefore well suited to countries like Great Britain which lack the sources of water power available elsewhere.

Sulphate of ammonia is sold as fine white crystals, is easily drilled, and stores well in dry conditions.

It is now more widely used in Great Britain than any other nitrogenous fertiliser.

In the soil ammonium sulphate undergoes two important changes.

(1) It reacts with the complex clay and humus constituents in the soil, giving rise to calcium sulphate and a complex ammonium body. The calcium sulphate being soluble washes out in the drainage water, but the ammonium complex is insoluble and does not, but remains held in the soil.

(2) The ammonium in the complex becomes nitrified by bacterial action, and presumably is changed to calcium nitrate through interaction with more calcium carbonate.

On chemical grounds one would expect that a dressing of 1 cwt. sulphate of ammonia would thus cause a loss of 1½ cwt. calcium carbonate from the soil.

Owing to the irregular distribution of calcium carbonate in the soil and to its irregular leaching by drainage water it is difficult to check these figures precisely in field experiments but analysis usually shows a smaller loss.

The discrepancy probably arises from a reappearance of some of the calcium as calcium carbonate in the soil; changes whereby this could come about are known.

Even if the loss of calcium carbonate does not amount to the full quantity expected from the equation, there still remains a loss of about 1 cwt. of calcium carbonate for each cwt. of sulphate of ammonia applied, and on soils deficient in lime this becomes very serious for two reasons: the lime is greatly needed for other purposes; and in its absence an acid residue is left in the soil, the ammonium or basic portion being more completely taken by plants than the acid portion. Now most agricultural plants will not tolerate this acidity, and in extreme cases completely refuse to grow. This remarkable action was first observed by Dr Wheeler at the Rhode Island Experiment Station in 1890,[1] and was investigated in an important series of experiments which showed that the trouble could be completely remedied by dressings of lime. A few years later the same phenomenon appeared at the Woburn Experimental Farm and was described by Dr Voelcker;[2] there also lime was found to be the proper remedy.

In practice this effect of sulphate of ammonia in making the soil acid is best obviated by a dressing of calcium carbonate somewhat greater than the total quantity of sulphate of ammonia applied during the rotation, which would frequently amount to 4 or 5 cwt. per acre. This is preferably given just before the barley in which the clover seed is sown.

Sulphate of ammonia is largely used for potatoes and may be given in quantities up to 4 or more cwt. per acre so long as other fertilisers are given in proportion. The tendency to acidity does not injure the potatoes, which are fairly tolerant of acid conditions, and it effectively checks the potato scab (*Actinomyces chromogenus* (*Oospora*)). Sulphate of ammonia is also much used for barley.

On grassland it acts very quickly, practically as quickly as nitrate of soda and it is about as effective. If

[1] Rhode Island Experiment Station, *3rd Annual Report*, 1891, p. 53; *4th Report*, et seq.

[2] *J. Roy. Agric. Soc.* 1897, p. 287; and subsequent years.

applied year after year on grassland it may considerably change the herbage because of its adverse effect on clover which becomes very pronounced unless lime, phosphate, and, if necessary, potash are also given. Also the soil acidity causes some species of grass to die out, thereby giving plants that can tolerate acidity a better chance of growth through reducing the competition for light, food and water. The effects are well seen on the Rothamsted plots: *Festuca ovina*, *Agrostis vulgaris*, *Anthoxanthum odoratum*, *Holcus lanatus*, *Arrhenatherum avenaceum*, *Conopodium denudatum* and *Rumex acetosa*[1] all tolerate the conditions set up by sulphate of ammonia: and so predominate on the plots that receive it continuously. The clovers, on the other hand, do not, and are soon suppressed.

This effect of sulphate of ammonia in controlling a mixed grass herbage is utilised in the making of lawns and golf greens. A mixture of 1 part of sulphate of ammonia with 2 parts of dry sand sprinkled on the lawn during dry weather at the rate of about 1 lb. per 5 sq. yd. rapidly kills daisies and other flat-growing weeds and also clovers. The addition of about $\frac{1}{3}$ part ferrous sulphate increases the effectiveness. Sulphate of ammonia, unlike nitrate of soda, is completely absorbed by the soil and shows no tendency to wash out. This can be demonstrated by packing on to a funnel 50 gm. of a heavy soil that has passed through the 100 mesh sieve, moistening with water where the soil touches the funnel, then gradually flooding with a $\frac{1}{2}$ per cent solution of ammonium sulphate. Collect the first 2 or 3 c.c. of the

[1] The ordinary names are sheep's fescue, bent grass, sweet vernal grass, Yorkshire fog, tall oat grass, earthnut or pignut, sorrel.

liquid that passes through, apply the Nessler test, and do the same for an equal quantity of the original solution. The two liquids show a striking difference in reaction to the test; that which has passed through the soil has lost almost all its ammonia but it now contains calcium, which can be detected by the ammonium oxalate test. Now repeat the experiment with a fresh lot of soil and a 1 per cent sodium nitrate solution. The nitrate shows no diminution in amount,[1] but some action nevertheless goes on and calcium occurs in the solution.

In consequence of this fixation in the soil, sulphate of ammonia is much in favour in tropical countries liable to torrential rainfall, and is used in the West Indies for the sugar cane and in India and Ceylon for tea, where its tendency to make the soil acid is an advantage.

It is not, however, necessarily better than nitrate of soda in wet seasons in temperate conditions such as those of Great Britain (p. 166).

Its fertiliser properties compare as on p. 174 with those of nitrate of soda.

When compared on the basis of equal nitrogen content nitrate of soda is frequently about 10 per cent more effective than sulphate of ammonia.

Commercial sulphate of ammonia is guaranteed to contain 20·6 per cent of nitrogen equivalent to 25 per cent ammonia: actually a higher percentage, 21·1 is usual: this is equivalent to 25·6 per cent ammonia. It is almost free from impurity: the chemically pure substance contains 21·2 per cent of nitrogen, equivalent to

[1] The brown ring test is suitable.

Nitrate of soda	Sulphate of ammonia
Ready for use by plant. Liable to loss by leaching.	Acts rapidly on crops other than grass only after it is nitrified: this may take 7–10 days or more.
Specially good in unfavourable seasons or in sour soil conditions.	Good in regions of torrential rainfall.
Conserves calcium carbonate in soil.	Uses up about its own weight of calcium carbonate.
Tends to keep soil neutral.	Tends to make soil acid; therefore lime must be added if necessary.
Sodium is specially useful for mangolds, sugar beet and on certain types of soil, but it tends to make heavy arable land sticky, though not grassland.	Sulphate has nutrient value for turnips, swedes, cabbage and other Brassicae, and it has no harmful physical effects on soil. Specially useful for potatoes, tea and, in irrigated areas, rice and oranges.
Contains 15½ to 16 per cent nitrogen.	Contains 21 per cent nitrogen and is now considerably cheaper per unit of nitrogen than nitrate of soda.

25·7 per cent ammonia. It is the most concentrated of all the nitrogenous manures in common use.

Calcium cyanamide. This fertiliser, like calcium nitrate, is made from air and limestone but the processes and the products are entirely different. Calcium cyanamide is not soluble in water: it is not a direct plant food: on the contrary, it is actually a poison to plants. In the soil, however, it undergoes a series of changes, first forming acid calcium cyanamide, then reacting with some soil constituent not yet recognised and breaking down to calcium carbonate and urea, and then, by the action of micro-organisms, to ammonium carbonate and calcium nitrate. The first changes appear to be rapid, taking only a few days in ordinary conditions; the conversion of ammonia to nitrate, however, is delayed even after all the cyanamide has decomposed. As it becomes converted to calcium carbonate in the soil it has no tendency to make the soil acid: indeed, as it contains free lime it acts the other way. Under bad storage conditions or in admixture with superphosphate a more complex compound, dicyanodiamide, may be formed, which is also harmful to plants and nitrifying organisms, and persists in the soil for much longer; it breaks down very slowly to form ammonia in the soil.

The special properties of cyanamide are its harmful effect on young vegetation, the delay in nitrification, and the possibility of its conversion into the harmful dicyanodiamide; in consequence it cannot be used in quite the same manner as nitrate of soda or sulphate of ammonia. Two precautions are highly desirable:

1. The cyanamide should be applied some days before the seed is sown, about 3 days for 1 cwt. per acre, 6 days for 2 cwt. per acre and so on in the same proportion.

2. It should be incorporated with the soil by harrowing.

In England its use is usually restricted to spring-sown crops. For spring-sown barley it has proved at least as effective as sulphate of ammonia. It has also given good results for sugar beet, especially in the western counties, for market-garden crops and grass orchards. It has not, however, proved as useful for potatoes. It is about 80 per cent as effective as sulphate of ammonia when compared on an equal nitrogen basis. It is generally used more in compound fertilisers than by itself.

Its direct harmful effect on vegetation makes it useful as a weed killer, and it has the further advantage that after it has killed the weeds it feeds the crop. Large quantities are thus used in Germany and Belgium, the material being applied in early spring as a top-dressing of about 2 cwt. per acre to cereals when the weeds have well opened their true leaves; cornflower, poppies, and charlock, all serious weeds in corn, are readily destroyed.[1] The corn itself suffers for a time but soon recovers, and the additional nitrogen and freedom from weed competition enable it to grow well. Good results have also been obtained in England.

Its usual nitrogen content is 20·6 per cent, this being adjusted to be the same as in sulphate of ammonia. Its total calcium is equivalent to 60 per cent calcium oxide; about a third of this is in the state of free lime.

Comparison of these nitrogenous fertilisers

As a general rule for all arable crops and soils examined, the relative effectiveness of the three fertilisers described above, when compared on the basis of equal nitrogen content is:

The nitrates	110
Sulphate of ammonia	100
Cyanamide	80

Comparative experiments on barley made at Rothamsted and its outside centres gave the following increases in cwt. of grain per acre for 0·2 cwt. of nitrogen (N) applied:

Sulphate of ammonia	2·8
Nitrate of soda	3·3
Nitrochalk	4·6
Cyanamide	2·6

[1] 1 cwt. per acre may suffice if it is applied with a dusting machine. Clover and grass seed should not be sown till several days after the application.

CHAPTER IX

THE PHOSPHATIC FERTILIZERS

THE effects of phosphate on the growing plant are very marked. In the seedling stage root growth is increased, leaf development is hastened and the first leaf stage is more rapidly passed; later on more tillers are produced by cereal plants and there is greater growth of shoots and, to a somewhat smaller extent, of roots. The number of shoots finally bearing ears also increases, and the ears emerge more rapidly on plants well supplied with phosphate, as compared with those that lack it. On soils suffering from phosphate starvation the ripening of barley is delayed; at Rothamsted the crop on the phosphate-starved plots is about 10 days later in maturing than that receiving superphosphate; the heads emerge later from their ensheathing leaves, and remain green long after the others are turning golden yellow. A consequence of this lateness of emergence is that the heads become more liable to destruction by the larvae of the gout fly (*Chlorops taeniopus* Meig.) which, hatching from their eggs on the topmost leaf, crawl downwards, seeking the young grain for food. Where superphosphate is given, the heads have usually emerged and grown upwards beyond the junction of the leaf and stem, and thus escape attack by the insects.

Soluble phosphatic fertilisers have a special action on leguminous crops. Easily soluble calcium phosphate facilitates or hastens a change of the nodule organism which enables it to move about in the soil, and so increases the chance of its getting into the roots of clover or lucerne and forming the nodules necessary for good growth. Phosphate is therefore used with great advantage in grassland for the encouragement of clover, especially wild white clover where this is able to grow. Phosphate starvation of plants is not so quickly recognised as nitrogen starvation. In extreme cases swedes,

Plot 1 3 5

Fig. 41. Effect of fertilisers on swedes. (Agdell field, Rothamsted, 1912.)
Plot 1. Complete manure—phosphates, potash and nitrogen compounds.
,, 3. Incomplete manure—phosphates, potash and no nitrogen compounds.
,, 5. No manure.

turnips and mangolds fail altogether; the leaves of cereals become dull greyish green in colour; while those of apple trees are bronze coloured and are carried mainly at the tips of the shoots.

The crops most sensitive to phosphate starvation are swedes, turnips and potatoes; then come mangolds, sugar beet, clover, barley, wheat and oats; grass is less sensitive.

Phosphate deficiency is common in many parts of the world, and it may occur on any type of soil, lowering the crop-producing power of soils in arable cultivation, and

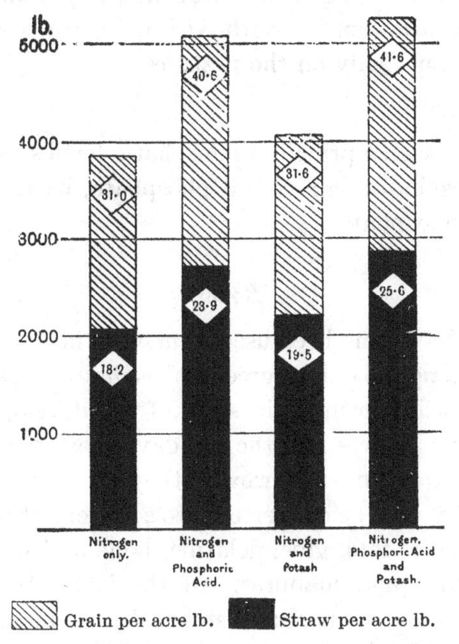

Fig. 42. Effect of phosphates and of potash on the yield of barley. (Hoos field, Rothamsted.) (Average 60 years, 1852–1911.)

The columns represent total produce per acre while the figures in the diamond spaces give bushels of grain and cwt. of straw per acre.

causing "deficiency diseases" in animals grazed on the natural herbage. Among the more important phosphate-deficient soils in England are the loams of the eastern counties, the acid Millstone Grits of the north and the oolitic soils of the west. Good examples of phosphate deficiency are seen on Hoos field at Rothamsted (barley), and on the experimental field at Saxmundham, East Suffolk (wheat, barley, mangolds, clover) and in Northern Ireland. The fen soils are very responsive to phosphate. Striking examples of the value of phosphatic manuring are seen in South Africa, Australia and New Zealand, especially on the pastures.

There are five principal phosphatic fertilisers: bones, superphosphate, ammonium phosphate, basic slag and mineral phosphate.

Bones

Bones have long been used as manure in England and were for long the only source of phosphate available; the demand still continues in spite of the development of alternative sources. In the old days they were used in the raw state: now, however, they go to the factory where they may be converted into a variety of products, bone articles, fats, glue, gelatine, bone and meat feeds for animals, and manures. Of the latter two are in general use: bone meal and steamed bone flour.

Bone meal. This is crushed bone from which fat has been removed but not the protein compounds. It commonly contains about 3·75 per cent of nitrogen and 20·5 per cent of phosphoric oxide (P_2O_5), the variation being as a rule 3·5–4·5 per cent of nitrogen and 20–25 per cent of P_2O_5. It is obtainable in various grades of fineness from powder to coarse particles. It is used for hops, fruit, market-garden crops, in gardens and as a base dressing in glasshouse practice. From the old experiments on farm crops in Little Hoos field at Rothamsted and also from the experiments at Saxmundham, it appeared that bone meal was about as effective as superphosphate, but there was no evidence of any special "lasting effects". No experiments on modern lines have, however, been made. Its popularity with gardeners probably depends on the fact that it cannot easily be misused.

Steamed bone flour. This is the bone manure most commonly used; it is made from bones after the fat and the nitrogenous constituents have been removed, the latter being used for making gelatine or glue. The residual bone is dried and finely ground.

It contains about 27·5–30 per cent of P_2O_5, this being higher than in bone meal because of the loss of nitrogen compounds. It is not altogether devoid of nitrogen, however, and still contains nearly 1 per cent.

No good recent field experiments have been made with steamed bone flour; in the old ones it was either inferior or else about equal to superphosphate.

But, like bone meal, it is popular with farmers and gardeners, and is used for grassland and root crops, and especially for adding to mixtures of fertilisers in order to improve their physical condition and their dryness.

Dissolved or vitriolised bones are made by treating bones with sufficient sulphuric acid to dissolve about half of the phosphate. In places they are used for potatoes and root crops and horticultural crops.

Table XVI gives the average composition of the various bone manures.

Table XVI. *Composition of bone manures*[1]

	Nitrogen	Equivalent to ammonia	P_2O_5	Equivalent to tricalcic phosphate
Raw bones	5	6	22	48
Bone meal	3·5–4·5	4·3–5·5	20–25	45–55
Steamed bone flour	0·8	1·0	27·5–30	60–65
Dissolved bones	3	3·6	15–16	33–35
Meat and bone meal	4–8	5–9	9–18	20–40

Meat and bone meal

This is prepared from tankage from the Argentine, from waste meat, condemned meat, refuse from slaughter houses, etc. The fat is first removed: it has commercial value but is useless as fertiliser. The residue is ground to varying degrees of fineness. The composition depends on the amount of bone present. Meat and bone meals are used for much the same purposes as bone meal.

Superphosphate

Even in the middle of the nineteenth century the supply of bones was inadequate to meet the demands of farmers: if no other source of phosphate had been discovered the world would have been faced with a difficult situation.

Fortunately, early in the nineteenth century, geologists discovered large deposits of calcium phosphate. These could not at the time be very finely ground, and so they had little if any fertiliser value. Lawes of Rothamsted, however, discovered the way of using them. Chemists

[1] In the old days bone ash was also used as fertilizer. It usually contained about 30–40 per cent of P_2O_5.

had shown that the phosphate became soluble if treated with sulphuric acid: Lawes worked out and patented in 1842 a method for doing this on the large scale and sold the product under the name "superphosphate". This was the first fertiliser to be made artificially; he thus founded the artificial fertiliser industry which now has grown to enormous dimensions.

Superphosphate is a mixture of soluble calcium phosphate and of calcium sulphate in about equal proportions. No separation is attempted, and the calcium sulphate is left in: it not only does no harm but has itself some fertilising value and indeed was much used in the past: it also serves to get the superphosphate into a dry condition because it absorbs water very completely. The process has attained a considerable degree of perfection, and allows of the production of a high-grade product, finely powdered or else granular, dry, and free from the defects of the older samples. The world's annual production now exceeds 15 million tons.

The rock phosphate comes largely from northern Africa and the United States: there are smaller deposits in the Pacific Islands.

It has been found convenient to standardise the various grades of superphosphate and sell them on a definite basis. The amount of soluble phosphate is determined by analysis as "phosphoric acid" (P_2O_5). There are two common grades in England now: 14 and 16 per cent.

A more concentrated superphosphate, often called "double" or "triple" superphosphate, is also available: it is prepared by treating phosphate rock with sulphuric acid as in the ordinary process, then washing out the phosphoric acid formed and using it to dissolve a further quantity of phosphate rock. The resulting product contains 42–46 per cent soluble P_2O_5. As it has been subjected to a double treatment it is called "double" superphosphate in Great Britain: but as it contains about three times the ordinary percentage of soluble P_2O_5 it is called "triple" superphosphate in the United States. Both names stand for the same thing; "concentrated" superphosphate would be a better term.

It is used in making concentrated compound fertilisers and in regions where carriage is an important consideration.

The phosphate in superphosphate is soluble in water but it is rapidly precipitated in the soil and only very small quantities are found in the drainage water: practically the whole of the superphosphate added to the Rothamsted soils during the past 60 years and not taken by the crop still remains in the soil. This can be shown by an experiment like that made with sulphate of ammonia (p. 172) but using a 5 per cent solution of superphosphate and testing the filtrate with ammonium molybdate. No precipitate forms; the original solution, however, gives first a yellow coloration and then after a short time a precipitate.

Superphosphate is sometimes described as an acid manure but the statement is incorrect: there is no evidence that it causes the soil to become acid: the Broadbalk plot manured annually with superphosphate for many years does not lose its lime any more quickly than the corresponding plot without superphosphate. A light soil at Woburn was dressed annually for 50 years with 3–3½ cwt. per acre of superphosphate: the reaction (pH value) remained the same as that of a neighbouring plot without superphosphate. It has no bad effect on the texture of the soil; on the contrary, it frequently causes an improvement.

Superphosphate should not usually be mixed with lime, though there are cases in the eastern counties where the mixture has given good results.

In spite of its solubility, however, plants take up only about 25–30 per cent of the phosphate given to them: the rest remains in the soil.

The ammonium phosphates

These combinations of ammonia and phosphoric acid are the most concentrated fertilisers available. Three are in use, generally known as A, B and C, with the following composition:

	A	B	C
Nitrogen (N)	13·8	17·9	15·6
Soluble phosphoric acid (P_2O_5)	41·4	17·9	31·2
Proportion	1 : 3	1 : 1	1 : 2

They are not used as ordinary fertilisers but for incorporation into concentrated compound fertilisers.

Basic slag

During the manufacture of steel it is necessary to eliminate the phosphorus which would otherwise be harmful: this is taken out in the basic slag. Prior to 1914 the Bessemer process was used, and was so worked as to give a slag containing 18–19 per cent P_2O_5, equivalent to 40–42 per cent tricalcic phosphate. Since then the open-hearth process has become much more common, and now supplies much, though not all, of the slag on the market. The samples vary in composition according to the conditions of manufacture but they are graded to uniformity by the slag grinders: the different grades go down to 8 per cent P_2O_5 (=17·5 per cent tricalcic phosphate). But a more important difference from the old Bessemer slag is that some of the phosphorus may be in combination with fluorine in which form it has little if any fertiliser value. The official test of solubility in citric acid should always be applied to basic slag, and only the soluble part should be reckoned as phosphatic fertiliser; though it may still happen that some of the other constituents, e.g. the lime, and the manganese are of value. In the normal good slags on the market 80 per cent of the total phosphate is soluble.

The phosphate constituents of the slag are complex in composition and are not expressible by any simple chemical formula.

Some free lime or calcium carbonate is usually present, and some of the calcium silicates in the slag can apparently neutralise soil acidity: 3 cwt. of slag has about the same lime value as 2 cwt. of calcium carbonate whether the slag is of high or low solubility in citric acid.

Basic slag is ground so that 80 per cent or more passes through a sieve having 100 meshes to the linear inch.

It has given remarkable results on clay grasslands, and has probably been the cause of greater improvement on these than any other single factor, its action being to bring on the wild white clover which then so increases the nitrogenous organic matter of the soil that greater growth of grass becomes possible. A usual dressing is 5 to 10 cwt. per acre, and it can be applied at any time during autumn or winter. A striking series of experiments was begun in 1897 at Cockle Park and has been continued ever since: these showed the improvement that could be effected in extreme cases. Not only is there a marked development of clover, but the mat of dead vegetation becomes decomposed: in consequence the rain water percolates through the soil more easily. The sour Boulder Clays are very responsive to slag, especially those sparsely covered with vegetation, with little or no clover but a certain amount of bent grass (*Agrostis vulgaris*). The clays overlying the limestones and the chalk have in many counties proved responsive also. On some of the very acid soils in the north of England it fails to act until lime is added, and on dry hill land it has also proved ineffective without potash, and even then the improvement did not repay the cost.

In recent years experiments made by the Rothamsted staff in different parts of Great Britain have shown the increased yields of hay given by slags and other phosphates. Some of these are set out in Table XVII showing the yields obtained from a single dressing applied in the winter of 1930–1.

Both yield and nitrogen content of the hay were increased, as also were its lime and its phosphate content. Thus the slag improves the feeding quality as well as the quantity of the herbage. There is no evidence that the low soluble slag improves in action as the years go by. It was no more effective in the second and later years than in the first, and while it might for several years give a steady return it gave no increasing return. The high-soluble slag was always superior to the low soluble: there was no sign of any catching up. Usually

Table XVII. *Northallerton hay experiment*, 1931–4

	Without phosphate	Low-soluble slag	Gafsa mineral phosphate	High-soluble slag	Super-phosphate	Standard error*
			Yield of dry hay, cwt. per acre			
1931	24·3	29·5	33·6	36·4	37·8	0·79
1932	12·8	15·2	18·6	19·3	18·7	0·43
1933	14·6	18·2	21·0	21·3	19·9	0·44
1934	9·5	12·0	14·0	14·3	13·2	0·16
Mean	15·3	18·7	21·8	22·8	22·4	—
			Nitrogen as percentage of dry hay			
1931	1·32	1·36	1·55	1·58	1·59	0·032
1932	1·56	1·68	1·82	1·81	1·89	0·063
1933	1·28	1·40	1·51	1·45	1·49	0·028
1934	1·24	1·32	1·34	1·36	1·36	0·022

* The "standard error" allows of an estimate of the part of the result attributable to errors, irregularities of soil, etc. and not to the treatment. A difference greater than three times the standard error can safely be accepted as due to the treatment: it is therefore called "significant". A difference less than this might be due to the treatment but it might not: one could not accept it as proving anything (see p. 271).

the crop takes up only about 25–30 per cent of the phosphate supplied in the high-soluble slag: the rest remains in the soil apparently in some inert form.

On arable soils high-soluble basic slag has a similar kind of action to superphosphate but is less effective; it promotes early growth of all plants, tillering of cereals, earlier emergence of the heads, and it increases final yields, though not usually as much as superphosphate containing an equal weight of phosphoric acid. Low-soluble slag has considerably less action. High-soluble slag is said to be better than superphosphate for roots on land infested with "finger and toe" but no good experiments have been made.

Mineral phosphate

Some natural phosphates when finely ground act quite well as fertilisers under certain conditions, and in recent years they have been increasingly used by farmers. About 10 per cent of the phosphate used in Great Britain and Ireland is mineral. Table XVII shows that Gafsa phosphate was nearly as good as any of the phosphates tested in that experiment: in other cases it has been inferior. As a rule the mineral phosphate appears to do best in moist, cool soils and in those tending to acidity. It should not be used in any quantity until actual trial has shown that it is effective. It may contain from 27 to 35 per cent of P_2O_5.

There are also certain treated phosphates on the market.

COMPARISON OF THE PHOSPHATIC FERTILISERS

Superphosphate is, in general, the most effective of the phosphatic fertilisers, especially in its first year. There is no evidence that it is any less "lasting" in its properties than slag or bone meal, nor that it increases soil acidity; on acid soils, however, it may prove less effective than a high-soluble slag.

Of the basic slags, the high-soluble is the most effective and is nearly up to the level of superphosphate in most cases. Slag of low solubility (below 40 per cent in 2 per cent citric acid solution) is less useful. Quite apart from its effect in enriching the soil in phosphate basic slag improves sour land by neutralising acidity; roughly speaking it has about two-thirds the lime value of an equal weight of ground limestone and sometimes more. The manganese present may be advantageous on soils deficient in manganese.

Mineral phosphate is very dependent on the conditions for its effectiveness. In some cases it is almost as good as high-soluble slag: usually, however, it was about equal to low-soluble slag. It acts best on acid soils, under high rainfall, and for perennial crops.

Bone meal and steam-bone flour are better for special than for ordinary use: they are comparable with basic slag in effect and in the tests hitherto made they have been about as good as superphosphate.

Crops vary much in their response to phosphatic fertilisers. Swedes and turnips come first, being most responsive, then potatoes and mangolds, then cereals in the order barley, oats, wheat; hay is less responsive in yield, except where the clover is much increased. Crops also vary in their power of taking up less-soluble phosphates; swedes and turnips have considerable power and can almost as readily assimilate the phosphate of basic slag, and sometimes even of mineral phosphate, as of superphosphate; potatoes, on the other hand, respond best to superphosphate.

Silicates. In cases of phosphate shortage soluble silicates have been used to make up the deficiency: they appear to bring into action some of the soil phosphates that would otherwise be unavailable. Where, however, phosphates are easily obtainable there is no advantage in using silicates.

CHAPTER X

POTASSIC FERTILISERS

THE system of agriculture long in vogue in this country consisted in selling grain and meat from the farm but returning the straw to the land in the form of manure. As the straw contains a large proportion of the potash

Fig. 43. Effect of potassic fertilisers on mangolds. (Barnfield, Rothamsted.)

Left-hand plot—Superphosphate and nitrogenous manure, no potassium salts. Right-hand plot—Superphosphate, nitrogenous manure and potassium salts.

while the grain and meat contain much phosphate, it is evident that the tendency of the system was to keep the potash on the farm and to reduce to a minimum the need for potassic fertilisers. Only milk and wool contain much potash: 1000 gallons of milk contain 15 to 20 lb. and 1000 lb. of wool contain about 50 to 60 lb. (as K_2O).

But as the area under potash-needing crops—mangolds, potatoes and sugar beet, etc.—extended, it became necessary to apply potash to the soil, and large quantities of potassium salts are produced in Germany, being mined in the region west and south-west of a line drawn from Leipzig to Brunswick; much also comes from Alsace. Deposits occur in Poland, Spain, Russia, the United States and elsewhere: potash also is extracted from the Dead Sea: all these sources are worked to some extent. It is present in wood ashes (these may contain 10 per cent K_2O), bonfire ashes, etc.

Potassium increases the efficiency of the leaf as a maker of carbohydrates from the carbon dioxide of the air: it is thus the counterpart of nitrogen which increases the leaf area but not the efficiency. On the Barnfield mangolds plots at Rothamsted potassic fertiliser does not greatly increase the weight of leaf either in the presence or in the absence of nitrogen, but in the presence of nitrogen it greatly increases the efficiency, causing a given weight of leaf to produce a larger amount of root (Table XVIII).

Potassium also keeps the leaves of plants green and functioning longer than they would otherwise do: this is of special importance on light soils but may be a disadvantage on heavy soils.

It is of special value for leguminous crops which fix gaseous nitrogen from the air.

It improves the health and vigour of the plant, enabling it to withstand adverse conditions of soil, climate or disease, etc. The plants well supplied with

Table XVIII. *Barnfield mangolds, Rothamsted: average annual yields, tons per acre for the* 50 *years* 1876–1928[1]

Nitrogen supplied	Other fertilisers super. +	Roots tons per acre	Leaves tons per acre	Roots yield per ton of leaf
Sulphate of ammonia containing 86 lb. nitrogen; also rape cake	Potassic fertiliser	22·5	5·20	4·33
	No potassic fertiliser	9·5	3·29	2·89
None	Potassic fertiliser	4·0	0·93	4·33
	No potassic fertiliser	4·5	1·05	4·28

potash at Rothamsted do better in bad years—whether of wetness or of drought—than the others: they are also more resistant to various diseases. On the Park grass plots supplied with excess of nitrogen but no potash the grass is liable to attacks of the fungus *Epichloe*, and in addition the seed heads are often barren; the mangolds get badly attacked by the fungus *Uromyces betae*; the wheat is subject to rust and the tips of the leaf begin to die early in the season and then the edges turn yellow for some distance down. Elsewhere also dressings of potash have enabled plants to withstand pests: flax liable to the "wilt" disease in Ireland, tomatoes in glasshouse culture, spring oats affected by eelworm have all benefited from dressings of potash. Potash manures also tend to counteract rankness of growth and therefore find valuable application for glasshouse and nursery work.

Further, potassium strengthens the straw of cereal crops and of the grasses: at Rothamsted the grass growing on the plots deficient in potash tend to become

[1] In two years the crop failed, and in one it was so poor that no nitrogen was given.

laid, especially in bad seasons. Finally, it improves the quality of fruit and the cooking quality of potatoes.

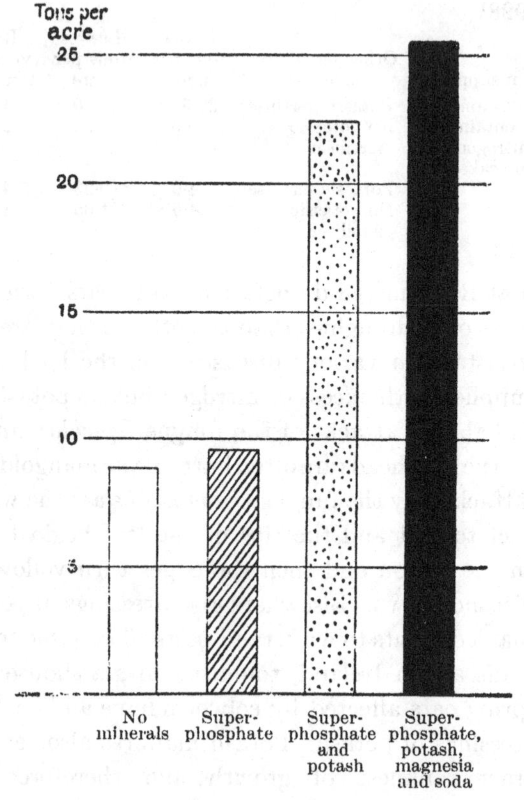

Fig. 44. Effect of mineral manures on the yield of mangolds already receiving nitrogenous manure (ammonium salts and rape cake), Barnfield, Rothamsted. (Roots, tons per acre, average 37 years, 1876–1912.)

The improvement is more marked on a light than on a heavy soil. Table XIX shows the marks assigned by competent cooks in Messrs Lyons' laboratories to steamed samples of Rothamsted and Woburn potatoes.

Table XIX. *Marks for the quality of steamed potatoes (Messrs Lyons' Laboratories)*

	Nitrogenous fertiliser			Potassic fertiliser	
Dose	Light soil Woburn	Heavy soil Rothamsted	Dose	Light soil Woburn	Heavy soil Rothamsted
0	34·4	29·2	0	32·6	28·5
1	33·3	29·3	1	33·6	29·5
2	32·9	29·1	2	34·5	29·6

On the light soil nitrogenous manuring has reduced the quality but potassic manuring has raised it; on the heavy soil the change has been less.

It is essential to maintain a proper balance between the nitrogen and the potassium in manuring. Frequently each acts better in presence of the other than it does by itself as is seen in Table XX.

Table XX. *Action of potash in reinforcing the effect of nitrogenous manure on potatoes*

	Increase given by sulphate of ammonia, tons per acre	
Mean yield tons per acre	Used alone	With potassic fertiliser
10·17	0·41	1·86

Potash is particularly needed for mangolds (Figs. 42 and 43) and potatoes; it increases both yield and the proportion of ware; a usual dressing is 2 cwt. of muriate or sulphate of potash for potatoes or 3 cwt. of 30 per cent potash salt for mangolds. It is of special value for fruit trees: deficiency is shown by a dying back of the tips of the shoots, by chlorosis of the leaf and dying at the edges of the leaf; Pershore plums show the chlorosis markedly: apples and bush fruit suffer from leaf scorch.

Potassium is often effective on grassland, especially on thin soils, and on leguminous crops. Other crops need it as the standard of farming rises and the yields are forced up: the natural supplies of potash in the soil are not always sufficient for the higher crops that ought to be obtained.

Light sandy soils respond considerably to dressings of potash, so also do moorland soils. It is because of the wide occurrence of these two types of soil in north Germany and of moorland soils in Sweden that potash is so much used in those countries; the demand is still further increased by the great quantities of potatoes and sugar-beet grown. Light chalky soils also respond to potash.

There are three potassic fertilisers in common use.

Sulphate of potash. Minimum guarantee 48·6 per cent K_2O. This is the most expensive of the potassic fertilisers, being made from the muriate, but it is the safest wherever quality is important. It should always be used for good-quality potatoes and for valuable market-garden crops.

It can be mixed with other fertilisers and applied at any time up to seeding. The usual rate is 1–2 cwt. per acre.

Muriate of potash. Minimum guarantee: potassium equivalent to 50·4 per cent K_2O. This is the most popular potassic fertiliser. It is practically as effective as the sulphate in increasing yields, but not quite so good in regard to quality of potatoes, though it can safely be used for potatoes that are sold straight off the field to customers not requiring the highest quality. It is apparently better than the sulphate for barley. It can be mixed with other fertilisers, though it is a little apt to cake: it can be applied at any time up to seeding. The usual rate is 1–2 cwt. per acre.

Unlike sulphate of potash, muriate of potash usually contains about 15 per cent of common salt.

Potash salts. There are two or three grades, but the most usual in this country is the one called 30 per cent potash salt because it is guaranteed to contain the equivalent of at least 30 per cent potash (K_2O). It contains also about 25 per cent of common salt (sodium chloride) and the German samples contain about 10 or 15 per cent magnesium salts, chiefly sulphate: the French samples, however, do not.

It is specially useful for crops that are benefited by common salt (see below), but should not be used for those which, like potatoes, are injured by it.

An older grade called kainit supplied potassium equivalent to 14 per cent of potash (K_2O), but this is no longer much used, being less economical than the 30 per cent potash salt.

AGRICULTURAL SALT AS FERTILISER

Ordinary salt or sodium chloride has special value for mangolds and sugar beet. Dressings of 5 cwt. salt per acre have added 4 tons of mangolds to the crop at Rothamsted, and at Woburn have given more sugar beet, containing a high percentage of sugar, and considerably more tops, even when muriate of potash is also supplied. On light soils salt is also beneficial on cereals and on grass. It economises the potassium in the plant, it enables the plant to obtain more potassium from the soil, and it apparently facilitates the uptake of water by the plant.

MAGNESIUM SALTS AS FERTILISERS

Deficiency of magnesium shows itself in a yellowing or "chlorosis" of the leaf, or in a premature loss of leaf. The best-known instance is the disease known as "sand drown", a chlorosis affecting tobacco in North Carolina. Apples and raspberries may, according to Wallace, lose their leaves too early through lack of magnesium if they receive no farmyard manure.

Magnesium is supplied by German potash salt and by farmyard manure, and deficiency is not common in Great Britain though it might occur on light soils.

MIXED OR COMPOUND FERTILISERS—UNIT PRICES

Since each of the plant foods exerts its own effect on the growing crop all are needed in order to obtain the maximum benefit. In general, all three of the foods, nitrogen, phosphorus and potassium, must be supplied during the course of the rotation.

This can be done in several ways. Each crop can receive one or more of the fertilisers described in the preceding pages, in quantities which experience or direct trial has shown to be best.

Alternatively a combined fertiliser can be used such as "potash super"; "potash mineral phosphate", which as their names imply are mixtures supplying both potash and phosphate. These are not definite chemical compounds like sulphate of potash, but mechanical mixtures: they are, however, made up to standard specification and should be uniform in composition.

Many farmers, however, find it more convenient to purchase their fertilisers ready mixed for each crop, and a number of so-called compound fertilisers are on the market, either in the powdered or the granular form. The best of these mixtures are made from the fertilisers described in the preceding pages, and those that include ammonium phosphate are more concentrated than those based on superphosphate.

The purchaser is legally entitled to a statement of the percentage of nitrogen (N); phosphoric acid (P_2O_5) soluble in water; insoluble phosphoric acid; and potash (K_2O).

The price of a compound fertiliser can be compared with that of a home-made mixture on the basis of the unit values. These are first worked out for the simple fertilisers. A unit is 1 per cent per ton; the unit value is the cost of 1 per cent per ton, and it is obtained by dividing the cost of the manure by the percentage of nitrogen, potash or phosphate. Thus the unit value of nitrogen in sulphate of ammonia was in 1939:

$$\frac{\text{Price per ton}}{\text{Percentage of nitrogen}} = \frac{£7.\ 14s.}{20{\cdot}6} = 7s.\ 6d.$$

The unit value of phosphate in superphosphate or in basic slag, and of potash in the various potash fertilisers, is obtained in the same way.

From the unit values the price of a home-made mixture supplying equal amounts of plant food to that in the compound fertiliser can be calculated. Some 10 or 15 per cent should, however, be added for mixing and other services, and items such as carriage, credit, etc., should be taken into account so that the comparison with the price asked for the compound fertiliser should be fair and entirely trustworthy.

All fertilisers should be stored in a dry place.

Mixing of manures

The following should *not* be mixed:

Sulphate of ammonia with calcium carbonate, potash salt,[1] lime, or basic slag.

Nitrochalk with potash salt,[1] super,[1] lime or basic slag.

Nitrate of soda with potash salt,[1] or super.[1]

Cyanamide with potash salt[1] or super.

Potash salt with any fertiliser unless the mixture is drilled at once.

Superphosphate with calcium carbonate, nitrochalk,[1] nitrate of soda,[1] cyanamide, potash salt,[1] basic slag or lime.[2]

Basic slag with sulphate of ammonia, super, nitrochalk or potash salt.[1]

Lime with sulphate of ammonia, nitrochalk, potash salt, or super.[2]

Calcium carbonate with sulphate of ammonia,[1] potash salt[1] or super.[1]

[1] Unless used at once. [2] But see p. 185.

CHAPTER XI

THE MINOR ELEMENTS OF PLANT NUTRITION

Boron. The need for boron was proved by Dr Katherine Warington at Rothamsted in 1923 (Fig. 45). In its absence sugar beet and swedes suffer from a "heart rot" or "crown rot"; potatoes from a disease rather like the leaf roll; apples in New Zealand suffer from "internal cork", and tobacco from "top rot". The trouble is intensified by over-liming and in dry conditions.

A dressing of 20 lb. per acre of borax is usually an effective cure. Even distribution, however, is essential; this can be effected either by mixing the borax with four or five times its weight of sand, so as to raise the total application to 1 cwt. per acre, or by arranging that the borax should be thoroughly mixed with the fertiliser.

Copper is essential to plant growth but in only very small quantities (Fig. 46). In practice copper deficiency is seen only on peat soils, or sandy soils from which peat has been removed or which are very poor in plant food. Striking examples are found in the Florida Everglades where on a calcareous peat soil marked increases in the yields of various market-garden and other crops followed the addition of 50 lb. of copper sulphate per acre. The trouble is well known in Holland and Germany, where on newly reclaimed peat or moorland soil, plants are liable to turn yellow and the tops of the leaves to become white and die; especially summer cereals, sugar beets and leguminous crops are affected.

Grassland farming on these soils proved almost impossible; in the second year after the grass had been sown all the grasses disappeared excepting Yorkshire

Fig. 45. Boron is essential for plant growth. The right-hand plants are supplied with culture solution formerly regarded as complete but containing no boron and therefore insufficient for plant growth. The other plants have received small quantities of boron and can now grow well. (K. Warington: experiments with broad beans.)

fog. Animals grazing on grass deficient in copper suffer from certain diseases; this trouble occurs in Australia also.

Fruit trees also suffer from copper deficiences. Citrus fruit trees grown on soils containing much organic matter suffer from the dying of the tips of the branches, the leaves and the fruits. The disease is called "die-back" or exanthema. A dressing of about 20–40 lb. copper sulphate per acre is usually a satisfactory remedy.

Fig. 46. Manganese, zinc and copper are all necessary for the growth of plants, though in very minute quantities. C. S. Piper's experiments with Algerian oats in water culture, 1938, Waite Institute, South Australia. All the plants receive culture solution, formerly regarded as sufficient, but unless the minute quantities of these elements are present growth cannot be complete.

Manganese. Lack of manganese in the soil leads to certain plant diseases, such as marsh spot in peas: these are curable by small dressings of soluble manganese salts.

Zinc. Fruit and nut trees in parts of the United States (California, Northern Central Florida and elsewhere) suffer from various physiological diseases;

Fig. 47. Small quantities of zinc are essential for the growth of plants. The small pine tree which Dr Kessell is showing has had no zinc sulphate, the large one to the right of the picture has received a small quantity.

mottling of leaf, "little leaf", or "rosette", and dying back of the growing tips. Ferrous sulphate was tried as a remedy, when it was found that some samples, used in large enough quantity, were effective, while others were not. Further work showed that the purer samples were least effective while others containing zinc sulphate as an impurity acted much better. Other workers observed that iron sulphate was successful in small-scale tests but not on the large-scale, and the difference was traced to the use of galvanised iron containers in the former but not in the latter. Zinc sulphate was therefore tried and it cured the disease. In parts of Western Australia pines suffer from zinc deficiency. About 20–40 lb. per acre of zinc sulphate is usually sufficient (Fig. 47).

Cobalt. In Australia and New Zealand certain obscure animal diseases, "pining" or "wasting" diseases, are attributed to deficiency of cobalt in the vegetation (Figs. 48, 49). There is the possibility of similar trouble in hill districts in this country.

Deficiency of iodine in the vegetation also causes trouble.

These elements are often called the "minor elements" in plant nutrition. Careful watch should be kept for deficiency symptoms, but the remedies should be applied only after expert advice has been obtained.

OTHER ELEMENTS

Sulphates. Sulphur is an essential constituent of plants, and occurs in some quantity in cabbages, swedes, turnips and other members of the Brassica family.

Lack of sulphur may cause plant diseases, e.g. "Tea Yellows" in Nyasaland. Both superphosphate and sulphate of ammonia supply sulphur to the crop, and sulphur deficiency is not known in this country. Sulphur appears to be particularly advantageous in dry regions.

Fig. 48. A "bush-sick" beast, State farm, Manakau, near Rotorua, New Zealand.

Fig. 49. The same land after the animals have received small quantities of iron salt, probably containing cobalt.

Chlorides. Though not essential to plants, chloride in small amounts appears to be beneficial; it enters the plant easily, and tends to keep the leaves and stems more turgid, greener, and less liable to lose water by transpiration. Mangolds, sugar beet and barley all benefit by the use of sodium chloride as a fertiliser (p. 197); potatoes, however, do not; indeed they suffer if large quantities are supplied.

Radioactive substances. From time to time radioactive substances are asserted to have special value, but there is no evidence that this is so.

CHAPTER XII

MANURES SUPPLYING ORGANIC MATTER: THE SIMPLER ORGANIC MANURES

THE older chemists made a rather sharp distinction between organic matter associated with living plants and animals, and containing carbon, hydrogen, oxygen and often nitrogen and other elements; and inorganic matter such as salts, air, water, usually much simpler in composition, and associated with minerals and other non-living substances. The distinction has now lost its sharpness though it is still retained as a matter of convenience.

The organic manures are those formed from plant and animal products and they differ from the fertilisers described in the preceding chapters in that they are much more complex, they are not themselves plant foods but have first to be decomposed in the soil. Some or all of their nitrogen is changed to ammonia and then to nitrate by the changes already described (p. 48): some of the phosphorus is changed to phosphate. The carbon, hydrogen and oxygen make no direct contribution to the plant food supply, but some of their compounds break down in the soil to form humus.

The value of organic manures thus depends on the amount of plant food and of humus that they can produce. These cannot entirely be determined by chemical analysis, and direct field experiment is always necessary. The production of plant food is complicated by the circumstance that the living organisms which bring it about must themselves feed on the organic matter, and some of them have the same requirements as plants. The nitrogen is particularly affected. For many of the usual organic manures, especially vegetable residues, the organisms consume about 1 part of nitrogen for each 30–40 parts of dry vegetable matter that they decompose. Organic matter containing about 1·5 per cent of nitrogen in the dry matter thus decomposes in the soil without production of nitrate, all the nitrogen going to feed the organisms and conferring no immediate advantage on plant growth. Organic matter richer in nitrogen has an excess of nitrogen over the requirements of the micro-organisms and this appears as nitrate, and contributes to the productiveness of the soil. But if the organic matter is poorer in nitrogen it may become positively detrimental to crop yield because in decomposing it the organisms take up from the soil ammonia and nitrate that would otherwise be available for the plants.

Thus cereal straw, which contains about 1·2 per cent nitrogen in its dry matter, generally slightly lowers the yield of crops when it is ploughed in, while if it is enriched with the excreta of animals and made into good farmyard manure containing 2–2·5 per cent of nitrogen in the dry matter it becomes very effective. While the value 1·5 per cent is not to be taken as a hard and fast limit, good field evidence should always be obtained before accepting claims of fertiliser value for substances containing only this quantity of nitrogen in their organic matter.[1]

[1] See p. 248 (Town Refuse).

Cellulose and lignin appear to be the main sources of humus; they occur in bulky plant materials like straw but not in animal products like meat, blood, etc.: these yield no humus. In any case only manures that are ploughed in at the rate of several tons per acre can add appreciably to the humus content of the soil.

Organic manures can thus be divided into those that furnish plant food only, those that furnish plant food and humus, and there is a third class of organic materials often proposed and sometimes used as manure that do neither, and are useless or actually harmful by inducing consumption of soil nitrates by the soil micro-organisms.

ORGANIC MANURES SUPPLYING PLANT FOOD ONLY

In the end the plant food produced by organic manures is exactly the same as that supplied by the fertilisers already described. But there is this important difference. The organic manures only slowly decompose and liberate the plant food, while the fertilisers supply it all at once.

The older agriculturists thought there was some special virtue in a slow-acting manure: it was supposed that plants, like animals, needed a daily ration of food, and it was therefore better that their food should be daily liberated rather than supplied all in one dose with the risk of loss. It is now known that this is incorrect. The plant takes up a great part of its necessary nitrogen, phosphorus and other mineral matter within the first few weeks of active growth: foods assimilated later may even do harm by reducing the quality of the crop. Nor is there any advantage in putting on a manure for the sake of an effect to be obtained in the following year. Quite apart from the risk of loss, 1 lb. of nitrogen yields the same amount of ammonia and no more, whether the decomposition process takes weeks or years: indeed, there is the disadvantage that a slow-acting manure represents capital locked up while the quick-acting manure gives a quicker return. It is sometimes asserted that organic manures supply to the crop something such as a vitamin or even more subtle substance that greatly enhances its nutritive or other value and that cannot be obtained in any other way. There is no evidence for these claims, and experiments so far made show that good healthy crops can be grown with the standard fertilisers only, and that their food value is the same as that of crops grown with farmyard manure. On present knowledge the best test of organic manures is the yield of crops they will give in the year of application and the one or two succeeding years.

The organic manures included in this chapter yield no appreciable quantity of humus, but may nevertheless act on the soil, affecting its physical condition or its reaction. This must of course be taken into account in determining their agricultural value.

A. *Manures of vegetable origin*

Oil cakes. The large demand for oils and fats has created an enormous industry in pressing out oil from oil seeds. In some cases, e.g. linseed and cotton, the residues have considerable value as cattle food: in other cases they are unsuitable for this purpose and are then offered as manure. The best known in this country is rape cake, which usually contains about 5 per cent of nitrogen, 2 per cent of P_2O_5 (i.e. 4 per cent of "phosphate") and 1 per cent of potash.

Rape cake has long been used as a manure with good results[1] in this country and in India (p. 50), and numerous experiments at Rothamsted and at Woburn have proved its value both on cereals and on roots (Tables XXI and XXII).

Table XXI. *Effects of rape cake and farmyard manure on barley and wheat, Woburn. Mean yields* (1890–1906), *cwt. per acre*

	Barley		Wheat	
Plot	Grain	Straw	Grain	Straw
10*b* (rape cake = 83 lb. N per acre)	17·7	20·6	14·6	25·1
11*b* (F.Y.M. = 105 lb. N per acre)	19·0	22·8	13·7	24·6

No comparisons have been made with farmyard manure supplying the same quantity of nitrogen: generally the farmyard manure is given in much larger amount. In the Woburn experiments the difference was less than usual, and here rape cake was more effective than farmyard manure per lb. of nitrogen supplied for wheat, though the difference was less for barley.

Rape cake proved inferior to sulphate of ammonia on market garden crops at Rothamsted and Woburn.

Its effect only lasts for one year at Rothamsted and no evidence has been obtained of any significant residual effect (Table XXII).

When used regularly year after year on the same land rape cake has the remarkable effect of causing the soil to become acid.

[1] An interesting old account is given by Hannam in *J. Roy. Agric. Soc.* 1843, 4, 177.

Table XXII. *Immediate and subsequent effects of rape cake. Rothamsted, Little Hoos field. Produce per acre*

	Wheat bushels	Barley bushels	Roots tons	Clover hay cwt.
Control, no rape cake	22·1	21·2	10·2	49·2
Additional yield for rape cake	5·0	12·6	1·6	−2·8
In 2nd year	0·4	3·6	0·4	1·2
In 3rd year	0·6	0·1	0·2	−0·9
In 4th year	0·1	−1·1	−0·6	−1·0

While rape and similar cakes have no special value in English agriculture there are circumstances in which they are very useful. In India, Japan and elsewhere rice is grown in swamp conditions, and here the standard fertilisers do not usually act well. Oil-seed cakes in these conditions are valuable fertilisers.

Malt culms. These are the rootlets of germinated barley removed during preparation of malt. They contain about 4 per cent nitrogen, 1·5 per cent P_2O_5 and 2 per cent K_2O.

In the few experiments made with them they were inferior to the equivalent mixture of artificial fertilisers.

B. *Manures of animal origin*

Guano. Guano consists of the droppings of marine birds mixed with feathers, corpses of dead birds, etc. It first came into this country from Peru in 1840 and rapidly achieved a high reputation. Other supplies have since been drawn from islands off the coast of South Africa, especially Ichaboe, but these are retained by the Union Government for South African consumption and are not generally shipped to Europe.

Other supplies amounting to 2000–5000 tons annually

come from the sandy coast of South-West Africa where artificial platforms on which the birds can nest have been erected in the lagoons. This guano contains many feathers. Smaller quantities are found on islands in the Persian Gulf, the Red Sea and off the coasts of north Somaliland and southern Arabia, mainly produced by gulls.

The composition and character of the guano depend on the initial material and the conditions under which it has accumulated. In rainless areas it rapidly dries and remains undecomposed; in wet areas it suffers considerable decomposition and loses much of its nitrogen and organic matter, becoming more phosphatic. Thus two grades of guano are available: the nitrogenous, obtained in rainless districts; and the phosphatic, from moister regions.

The chief supplies of high-grade nitrogenous guano are from the rainless islands off the coast of Peru: a small quantity is produced in northern Chile, but this is reserved for Chilean agriculture. The birds[1] are carefully preserved and the removal of the guano is done systematically during the non-breeding season on a two-yearly rotation, about half the islands being worked each year. Some 150,000–160,000 tons of fresh guano are thus obtained per annum, though in recent years it has been retained for Peruvian use. Improvements have been effected in the technique of the industry, and in consequence the nitrogen content is now higher than it used

[1] The birds chiefly concerned are the guanay (*Phalacrocorax bougainvillei*), a cormorant peculiar to the Humboldt Current but of antarctic affinities; the camanay (*Sula nebouxi*); the alcatraz (*Pelecanus thagus*); and the piquero (*Sula variegata*). (See R. C. Murphy, *Bird islands of Peru*, Putnam, 1925.)

to be: 10 per cent was a frequent value, especially where the deposits were very old, but now the high-grade samples average 14–16 per cent.

The phosphatic guanos are obtained chiefly from the north of Peru and from the Seychelle Islands.

The percentage composition of these guanos is given in Table XXIII.

Table XXIII. *Percentage composition of the various guanos*

	Nitrogen	P_2O_5	Equivalent to tricalcic phosphate	Potash K_2O	Moisture	Organic matter	Sand
Nitrogenous							
High-grade Peruvian	13–16	9–11	20–24	2–3·5	18–28	50–60	1–5
S.W. African guano	11–15	10–15	20–26	1–2	15–20	—	—
Phosphatic							
Low-grade Peruvian	1·25–2·5	15–25	35–55	1–1·5	10–20	6–10	40–45
Seychelles	0–1	25–30	56–65	—	15–20	—	—

In this country guanos are used mainly for horticulture. The old high-grade guano (11·0 per cent nitrogen) was tested at Rothamsted during the period 1904–16; it was quick acting but its effect lasted only for one season (Table XXIV); it was apparently superior to

Table XXIV. *Effect of Peruvian guano on yield of crops. Rothamsted, Little Hoos field. Produce per acre*

	Wheat bushels	Barley bushels	Roots tons	Clover hay cwt.
Control, no guano	22·1	21·2	10·2	49·2
Additional yield for guano	6·2	17·7	2·9	0·5
In 2nd year	−0·3	−0·4	0·3	−2·6
In 3rd year	−0·7	−3·0	0·4	2·1
In 4th year	−0·1	−1·3	−0·5	1·2

rape cake and shoddy. In the only recent experiment recorded high-grade guano was stated to be inferior to the mixture of standard fertilisers, including sulphate of ammonia.[1]

Poultry manure. The great increase in the numbers of poultry in Great Britain has led to an increase in amount of poultry manure available. It can be used fresh but it can also be dried and stored.

Samples recently analysed at Rothamsted had 88 per cent of dry matter containing 3·8 per cent of nitrogen, 3·5 per cent P_2O_5, 1·7 per cent K_2O and 50–60 per cent of organic matter. In the year of application 100 parts of its nitrogen was equal to about 65–70 parts of nitrogen from sulphate of ammonia, i.e. 8 cwt. of the poultry manure had about the same value as 1 cwt. sulphate of ammonia. Some improvement results if poultry manure is used again on the same soil, i.e. there is a small residual effect, but it still does not catch up to sulphate of ammonia.

Manufactured manures

Fish guano or manure. Fish guano is obtained from the fish offals or unsold fish from various markets throughout the country. The livers are first extracted for subsequent treatment and the remaining material is heated in large steam-jacketed cookers, the arms of which rotate on a centre piece while the vapours are removed by fans. If the fish is fresh the product is used for feeding and is called fish meal; its oil content is reduced to approximately 5 per cent; the nitrogen content is about

[1] *Rothamsted Annual Report*, 1917, for the old experiments; Dent, Manson and Trevains, *Empire J. Exp. Agric.* 1938, **6**, 176, for the recent experiments.

11 per cent (=70 per cent albuminoids) and the phosphoric acid (P_2O_5) content is about 8 per cent.

But if the fish is old, or taken from rubbish, it is converted into fish guano. The oil content is reduced by a solvent process to 2 per cent and the residue is ground; it contains about 7–9 per cent nitrogen; 3–8 per cent phosphoric acid[1] (P_2O_5); and 1 per cent K_2O. It is used for market garden crops, for flowers, hops, in private gardens, and at the rate of 4 lb. per rod on lawns on thin dry soils. It is also useful in farm practice. In the Rothamsted and other experiments it has usually been practically as effective as the equivalent mixture of artificial fertilisers and like them it left no residual effect. It must be worked into the land quickly or it may be taken by birds.

Dried blood. This usually contains about 12–14 per cent of nitrogen, though occasional samples rise nearly to 15 per cent. It is the quickest acting of all organic manures and hence it commands a specially high price. It is used in high-class horticultural work, e.g. for roses, carnations, vines, etc., and much of it is bought for America and for the better grade of mixed and patent fertilisers. In the Rothamsted potato experiments it was not superior to sulphate of ammonia used in equivalent amount.

Hoofs and horns. Good samples of these contain 12–14·5 per cent of nitrogen. They are sometimes used separately, sometimes mixed: and they may be finely or

[1] A large number of samples from the North Shields district analysed by Collins had the following mean composition: N, $8{\cdot}0 \pm 0{\cdot}2$ per cent; P_2O_5, $5{\cdot}9 \pm 0{\cdot}8$ per cent; K_2O, $1{\cdot}1 \pm 0{\cdot}3$ per cent. Samples offered in the Lea Valley had had some potash added, and contained: N, 7·7 per cent; P_2O_5, 8·5 per cent; K_2O, 4 per cent. Some samples contain much oil, 10 or even 20 per cent.

coarsely ground. They are slower in action than blood, and are much used in glasshouse work, especially where a shortage of nitrogen in mid-season has to be provided against. Hoof meal is preferred to horn in practice.

Seaweed

Seaweed is one of the oldest manures known and has been in use since remote ages in the coastal districts of Great Britain. It is so important in Jersey that the dates for cutting are annually fixed and announced by the Court, while the collection, drying and stacking afford regular summer occupation to some of the poorer people: in Scotland also the right to collect it sometimes forms part of the covenant with the landlord. Seaweed contains about the same amount of nitrogen as ordinary farm crops, and a considerably higher percentage of potash than these or the *Zostera* and other plants often collected with it. The different weeds vary, the long broad leaf-like *Laminaria* being richer than *Fucus*, the common bladder-wrack of the rocks. Further, the weed cut or thrown up early in the year is richer than that obtained later in summer or autumn. The composition of wet weed is usually:

Water	Organic matter	Nitrogen	Potash (K_2O)	P_2O_5
70–80	13–25	0·3–0·5	0·8–1·8	0·02–0·17

It is thus somewhat inferior to dung, containing less nitrogen and phosphoric acid (p. 233).

It is largely used for potatoes in Jersey and in Scotland, the dressings per acre being from 25 to 30 tons in Ayrshire and up to 45 tons in Jersey: some artificials are also applied. In Thanet 10–15 tons per acre are applied to lucerne and to market-garden crops.

Shoddy

A certain amount of waste material arises during the working up of wool in the mills, and this is available as manure under the general name shoddy. Some of it is beaten out of the fleeces: this includes the short wool, sand, burrs and other vegetable matter; some is short wool which falls out during combing, spinning or weaving of the wool or reworking of material from woollen rags; while some comes from carpet waste, rag cuttings, etc. A certain amount of oil, sud cake or other substances may also be present. Shoddy is thus variable in composition, but so far as is known its manurial value depends only on the wool present.

Pure wool contains about 17 per cent of nitrogen, and the various other substances present in shoddy contain little or none. Hence the nitrogen content affords the easiest index of the manurial value of the shoddy. High-class samples may contain as much as 14 per cent of nitrogen; low-grade material may contain only about 3 per cent.

The better quality samples are of considerable manurial value and are used for hops, fruit and market-garden produce. The low grades should be used only when they are known to be effective.

At Rothamsted a dressing of 1 ton per acre of a shoddy (flock dust) containing 12·5 per cent of nitrogen gave a marked increase of crop in the year of application and sometimes in the year after, showing that the material may have more after-effect than some of the other organic manures.

Table XXV. *Effect of shoddy on crops. Rothamsted, Little Hoos field. Produce per acre*

	Wheat bushels	Barley bushels	Roots tons	Clover hay cwt.
Control, no shoddy	22·1	21·2	10·2	49·2
Additional yield for shoddy	3·5	12·1	2·6	2·3
In 2nd year	4·4	−0·8	2·1	−4·2
In 3rd year	2·8	0·5	0·6	−0·6
In 4th year	0·4	5·0	−0·6	−5·4

No experiments have been made to find out whether shoddy has any special effect on the soil apart from supplying nitrogen.

Other waste products[1]

From time to time various nitrogenous or phosphatic substances are available as manure and can be purchased at fairly cheap rates. Their value depends on their composition and their mechanical condition: they should, therefore, be purchased only after analysis. The proper way of dealing with them would be to submit them to a preparatory grinding and mixing, but often the supplies are too small or too irregular to justify the erection of plant for the purpose.

Hair, calf hair, etc. contains about 10 per cent of nitrogen but is very slow to decompose in the soil especially in its usual long state: it should only be used when it can be obtained very cheaply and is in fair mechanical condition.

Feathers containing about 9 per cent of nitrogen are used with advantage in hop gardens, the small ones especially decomposing fairly quickly. On the other

[1] Described in more detail in Board of Agriculture Leaflet no. 175.

hand large feathers only break down slowly, the shafts especially taking a long time to decay.

Rabbit waste consists of the ears, feet, tail, etc., of the rabbit, and so far as the supply goes it is distinctly useful as manure and is improved by properly grinding.

Leather waste. Boot leather waste has at times been offered to farmers, but it has never been shown to possess manurial value and should not be purchased, nor should it enter into the composition of a mixed manure.

Soft leather scraps obtained from glove factories are in a different category, and find valuable application in market gardens in the glove districts of Worcestershire: they are put into the soil with young sprouts, cabbages, etc., at the time of setting out and afford a useful root run.

There is the possibility that leather treated by certain chemicals may be useful as fertiliser.

Soot and flue dusts. Several types of soot and flue dusts are available, only two of which, however, possess manurial value.

1. Household soot, which may contain anything from 2·5 to 11 per cent of nitrogen (equivalent to 3–13 per cent of ammonia); the lower values are given by mixed samples from dwelling houses, and the higher ones from the light fluffy soot, obtainable from sitting-room or kitchen chimneys; a usual amount may be put at about 4 per cent. Where soot is purchased by the bushel it is useful to know the quantity of nitrogen per bushel; this is much less variable than the figures for the percentage composition, because the rich soot is very bulky and the poor soot is more dense. Analyses at Cambridge showed that as a rule 1 bushel of soot con-

tains 1 lb. of nitrogen (mainly as sulphate of ammonia) normally worth about 4*d*. It contains some substance disagreeable to slugs and other pests, and is considered to improve the physical conditions in the soil partly by ameliorating the texture and partly by the warming effect of its black colour. It is applied at the rate of 20–30 bushels per acre as a spring dressing in market gardens, supplying both the nitrogen and the warmth that is then needed; and it is also used for hops in Kent, quantities being sent for the purpose from Manchester and other northern cities. It is probably about as effective as a dressing of sulphate of ammonia supplying an equal amount of nitrogen.

2. Blast furnace flue dust. This is quite a different substance from the foregoing, containing little or no nitrogen but some potash which has been volatilised from the coal, ore and flux by the intense heat of the furnace.

It has been used as fertiliser when the standard fertilisers were not obtainable; it is, however, liable to contain harmful substances such as cyanides, sulphocyanides, etc., and so should be purchased only on the advice of a competent expert.

3. Soot from destructors and boilers. This has little or no value, as it commonly contains only about 0·5 per cent of nitrogen, and less than 1 per cent of potash. On no account should any of these be purchased except on the recommendation of a disinterested analyst.

Sewage and sewage sludge

The daily consumption of protein per head of population in Great Britain is estimated at 87 gm., i.e. just over 3 oz., of which 46 gm. comes from animal and 41 from vegetable sources.[1]

In the course of a year this protein has supplied 11 lb. of nitrogen most of which is excreted, and a considerable part appears in sewage. Adequate data for estimating the amounts of phosphates and potash do not exist, but assuming they were each about one-quarter the value of the nitrogen the fertiliser value per annum of the excrements of the population of the United Kingdom (44·5 million) would be:

	Nitrogen	P_2O_5	K_2O	Value per annum*
Total, tons per annum	219,000	55,000	55,000	£10,434,000
Average lb. per head per annum	11	$2\frac{3}{4}$	$2\frac{3}{4}$	4*s*. 8*d*.

* When sulphate of ammonia = £7. 14*s*. 0*d*. per ton; 14 per cent super = £2. 16*s*. 0*d*.; and sulphate of potash = £9. 11*s*. 0*d*.

This is considerably greater than the value of the fertilisers used, which is roughly about £6 million. Owing to various losses, however, only a small part of this ever reaches the sewage works. An average quantity of domestic sewage is 25 gallons per head per day, containing 10 parts of nitrogen per 100,000; this corresponds to 9 lb. of nitrogen per head per annum.[2] Further, only a portion of the population is connected with sewage systems. Unfortunately no practicable means of realising the full value of sewage has yet been devised. Broad irrigation and sewage farming answer under certain conditions, but not as general methods of treatment.

[1] J. B. Orr. *Food, Health and Income*, London, Macmillan, 1936.

[2] Figures for certain towns in Germany indicate that the crude sewage per head per annum contains 8–14 lb. nitrogen, 1·8–3·4 lb. P_2O_5, and 5–7 lb. K_2O. The estimate for K_2O seems very high.

The only material generally available is the sludge which is prepared by some precipitating or settling process, and therefore contains only the insoluble compounds and not the soluble and valuable nitrates, ammonia, etc. This is its weakness: it has been so well washed during the process of formation that it has lost much of its decomposable material.

Various experiments have frequently been made to ascertain the manurial value of sludge, but the results have not been very satisfactory. The usual course of events is that farmers are first induced to purchase it but finally have to be paid to take it away.

Yet it has manurial value. The fresh lagoon sludge contains somewhere about the same amounts of nitrogen and phosphoric acid as farmyard manure, but less potash (Table XXVI). When used at the same rate as farmyard manure (about 12 tons per acre) it gave in Cranfield's[1] experiments somewhat similar results, except that it did not decompose in the soil quite so easily.

Table XXVI. *Composition of lagoon sludge* (*H. T. Cranfield*)

	Majority of samples %	12 typical samples	Farmyard manure
Moisture	50–80	70	76
Organic matter	10–30	19	—
Nitrogen	0·5–1·0	0·78	0·6
Phosphoric acid	0·1–0·5	0·31	0·25
Potash	0·03–0·3	0·12	0·6

So long as the sludge is within carting distance and can be applied at the same rates as farmyard manure it appears to be definitely useful.

[1] H. T. Cranfield, *Empire J. Exp. Agric.* 1939.

The problem becomes much more difficult when the local farmers cannot take up all the available material. Some method of drying has then to be adopted in order to give a more concentrated manure which can better bear the charges of transport. Considerable difficulties have arisen. Pressing the sludge brought down the water content to about 50 per cent but gave a hard cake of little manurial value. Other methods have been devised for improving the sludge. In some places, e.g. Glasgow, Kingston, etc., other materials are added to enrich it. These sludges are sent out in good mechanical condition ready for distribution: some of them at prices considerably in excess of their real value. At Bradford and Huddersfield a process is at work to extract the fat, grease, etc., which in modern times have become too precious to lose even in sewage; the resulting products contain respectively 2 and 3½ per cent of nitrogen and are of distinct fertiliser value.

Activated sludge obtained by blowing air through the sewage is much richer than the usual material; it may contain as much as 6 per cent or more of nitrogen and over 4 per cent of P_2O_5. It has good manurial value.

SUMMARY

The best of these organic manures is about as good as the equivalent mixture of the standard fertilisers, and no clear evidence has been obtained that any of them is better. They frequently command a higher price per unit of plant food, especially for horticultural purposes.

Probably the reason is that organic manures can be handled by gardeners more safely than the ordinary fertilisers; an excess does less harm, and there is less fear of scorching. Where cost does not enter into consideration they can safely be used, but where it does, experiments should be made to ensure that the extra expenditure on them is justified.

The experiments on agricultural crops are not numerous but their results all agree. This is true also for fruit and other horticultural crops. Many experiments have been made in the United States on apple trees: in these the standard artificial fertilisers came out better than the organic manures, especially under systems of low cultivation where nitrogen is most required. In the Long Ashton experiments on apples under clean cultivation no differences have been observed between the organic and inorganic fertilisers.[1]

Other claims have been put forward for organic manures. It has been asserted that crops grown with them are more healthy, and are more nutritious and healthful as foods, than crops grown with inorganic fertilisers. There is no evidence at all for these statements. Wheat and barley grown at Rothamsted on land which for over 80 years had received artificial manures alone and no organic manure, showed no superiority in content of vitamin B over that grown on land which for the same period had received farmyard manure annually.

[1] T. Wallace, "Manuring of fruit plantations and orchards", *J. Roy. Agric. Soc.* 1931, **92**, 125–142.

CHAPTER XIII

ORGANIC MANURES SUPPLYING PLANT FOOD AND HUMUS. FARMYARD MANURE, COMPOSTS, TOWN REFUSE

FARMYARD manure consists of the solid and liquid excretions from the animals together with the litter. It is the oldest and the commonest of all manures; indeed, in the old days beasts were sometimes kept on the farm solely for the value of the manure they made, and the practice still continues.

About half of the bulky food supplied to the animal (hay, straw, etc.) and nearly all the concentrated food (corn, cake, etc.) can be broken down by the digestive fluids in its body; the remainder cannot, and simply passes out as solid excreta or faeces. The digested portion enters the circulation and is used by the animals, most of the nitrogen and potash then finds its way into the urine. The compounds in the urine thus represent the easily decomposed part of the food, and in the soil they readily change to ammonia and other useful substances. On the other hand, the solid excreta, which could not be broken down in the body, prove somewhat resistant in the soil. Hence the urine is the most valuable part of the manure.

The richest manure is therefore that which contains the most and the richest urine. The value of the urine depends on the food, since it gathers up most of the digested nitrogen; the more digestible nitrogen the food contains, the better the manure produced. Concentrated foods like cake, containing much digestible nitrogen, therefore improve the dung. It does not follow that the most expensive cake gives the richest manure: the cost of cake depends on the oil present, while the value of the manure depends on the nitrogen. A linseed cake containing 7 per cent of oil gives better manure than a more costly cake containing 10 per cent, and decorticated cotton cake gives a richer manure still.

The manurial value of the urine also depends on the animal. Fatting animals keep back very little of their nitrogen—only about 5 per cent—and pass most of it out in the urine. Growing animals and milch cows keep back considerably more, so that the urine is correspondingly poorer. Consequently fatting animals make better manure than young stock or dairy cows.

Since the urine contains most of the potash and more than half of the nitrogen it should not be allowed to waste: sufficient suitable litter must be added to absorb it all. Straw, peat moss, and bracken are used for the purpose; they not only absorb the urine but also enrich the dung because they themselves contain valuable fertilising materials.

Straw is the commonest form of litter: it contains a fair amount of nitrogen and of potash, it has considerable power of absorbing urine, and it encourages a biological fixation of ammonia. Its composition varies somewhat (Table XXVII), but on an average 1 ton contains some 6*s*. worth of fertilising material.

Bracken compares favourably with straw on heavy soils: on sandy soils, however, it suffers from the drawback that it decomposes more slowly.

Table XXVII. *Typical analyses of the materials used for litter*

	100 lb. of each material contain		
	Nitrogen	Phosphoric acid, P_2O_5	Potash, K_2O
Oat straw	0·50	0·24	1·00
Wheat straw	0·45	0·24	0·80
Barley straw	0·40	0·18	1·00
Bracken	1·4	0·2	0·1
Peat moss	0·8	0·1	0·2

Peat moss has a higher absorbent power than straw and so gives a manure richer in ammonia than ordinary manure. On the other hand, it does not itself contribute as much to the manure as straw, being poorer in potash and phosphoric acid, nor does it decompose so readily; it is therefore less useful on light soils.

The manure as made. Knowing the weight and composition of the food and litter and deducting the food constituents retained by the animal, it is easy to calculate the amount of fertilising materials that should be present in any particular lot of farmyard manure. Numerous experiments show that the calculation does not come out right, the quantity of nitrogen found in the manure being usually about 15 per cent less than was anticipated. The loss does not take place in the animal: it is due to the action of micro-organisms while the manure is in the stall and before it is removed.

In view of the great variability in the quantity and composition of the litter and of the food it is obvious that no very definite figures can be given for the composition of farmyard manure. Numerous analyses have been made; a few are given in Table XXVIII.

The figures show that, on the average here given, a 10 ton dressing of farmyard manure supplies 135 lb. nitrogen, 55 lb. phosphoric acid, and 180 lb. potash; amounts present in a mixture of 640 lb. sulphate of ammonia, 400 lb. 14 per cent superphosphate and 360 lb. muriate of potash. As will be shown later, however, the nitrogen of farmyard manure is much less effective than in the sulphate of ammonia.

Changes on storing. Dung cannot generally be used directly it is made but often has to be kept for a period and applied to the land when convenient.

During this time it is undergoing constant change. Bacteria, moulds, etc., bring about decompositions of the kind already described (p. 46): they convert some of the nitrogen compounds into ammonia but they also assimilate ammonia: they oxidise some of the carbohydrates to carbon dioxide and water and they produce humus.

The amount of change depends on the amount of air that can get into the heap and on the extent to which it is washed by rain.

Air supply. The decomposition goes far if the heap is loose so that air can get in, and if it is occasionally turned. The nitrogen compounds break down and a good deal of the ammonia volatilises; there may be other losses of nitrogen also. The carbohydrates break down to various products including carbon dioxide, humus, etc. Heat is generated and the temperature may rise considerably, especially if the manure is rather dry, e.g. horse dung: wetter manure like cow dung does not become so hot.

For special purposes gardeners sometimes prefer a

well-turned well-rotted manure, made with free access of air. But for ordinary farm purposes this is too costly: the loss in making is too great.

The changes and the losses are both reduced if the heap is well compacted by the trampling of animals or by taking a cart over it so that air does not so easily penetrate. Changes still go on, it being impossible to exclude air completely, but they are less and so far as they go they are mainly good.

Water supply. If the heap becomes too dry moulds develop and take up the ammoniacal and amide nitrogen leaving much less available nitrogen for the crop. Further, the straw does not rot down, and when it gets into the soil its decomposition may involve the assimilation of soil nitrates by the micro-organisms concerned.

But the manure always loses a great deal if it becomes too wet, particularly if the heap is exposed to much rain. Much dry matter and nitrogen are lost; the black liquid that trickles away from the heap simply shows that losses are going on but gives no indication of their seriousness.

Thus the changes are greatest in heaps to which air is admitted: they are least when air is excluded. The losses vary in the same way. They are least when air is excluded and the heap is sheltered from the rain, i.e. in anaerobic conditions, in a compact heap stored under cover, or, what comes to the same thing, in manure made in a box and kept under the animal. They become greatest—amounting to 40 per cent or more—when the manure is made in open yards and then loosely packed into heaps and exposed to rain in the open. In the

Rothamsted experiments the losses on storage for three months were:

Compact heap under cover:	4 per cent of nitrogen.
Loose heap under cover:	7 " " " "
Heap exposed to open:	33 " " " "

The loss fell mainly on the ammonia and amides, i.e. on the easily available nitrogen (Fig. 50).

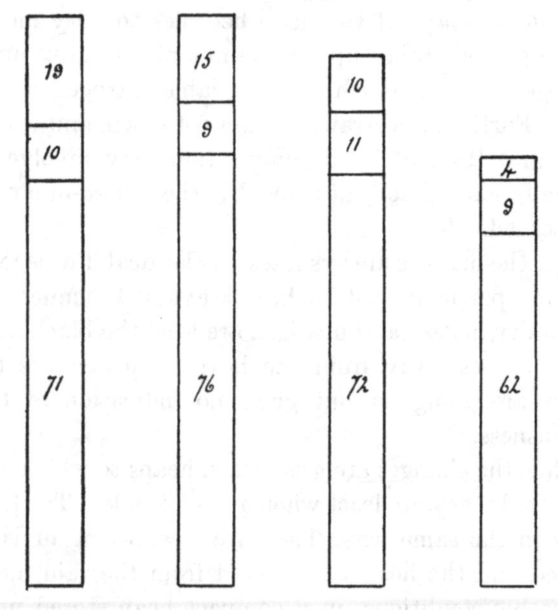

Fig. 50. Changes in nitrogen compounds in farmyard manure (cow manure) kept for three months, January 23 to April 30, 1914.

On this basis a 100 ton heap of manure valued at 10*s*. a ton would have lost over £16 worth of material in three months' exposure to the weather.

Formerly it was supposed that the loss of nitrogen took place mainly as ammonia, and farmers were advised to mix superphosphate, gypsum, or soil with the heap as "fixers", but these have not proved useful. Shelter and compacting seem the best methods of reducing the loss.

Table XXVIII. *Percentage composition of farmyard manure*

	Average 14 samples Rothamsted 1931–5	Effect of cake feeding		Dairy cows	Horse dung
		Cake	No cake		
Dry matter	25·0	27·4	27·2	19·4	26·9
Organic matter	20·2	21·8	21·6	15·2	21·9
Ash	4·8	5·6	5·6	4·2	5·0
Nitrogen	0·61	0·77	0·54	0·43	0·54
Phosphoric acid	0·26	0·39	0·23	0·19	0·23
Potash	0·80	0·60	0·67	0·44	0·54
As lb. per ton:					
Nitrogen	13·5	17	12	9·5	12
Phosphoric acid	5·5	9	5	4·5	5
Potash	18	13·5	15	9·5	12
Expressed as percentage of the organic matter:					
N	3·0	3·5	2·5	2·8	2·5
P_2O_5	1·3	1·8	1·1	1·25	1·0
K_2O	4·0	2·75	3·1	2·9	2·5

The following are the general rules for the more efficient conservation of the fertilising constituents of farmyard manure:

(1) *Manure made from fatting beasts.* If this is made in covered yards it should be left under the beasts until it is wanted. If made in open yards the manure should as soon as convenient be hauled out and tightly clamped.

The clamp should be so placed that it is not unduly exposed to rain. If any black liquid is running away it is a sign that shelter is insufficient and that wastage is going on. It is not sufficient to collect the liquid and pour it back again, steps should be taken to provide more shelter also. The clamp should not be disturbed until it is wanted.

If it is possible to avoid making the clamp by putting the manure straight on to the land and ploughing it in, so much the better. As far as possible summer storage of manure should be avoided.

(2) *Manure made from dairy cattle.* This has usually to be thrown out daily. It should be well protected from rain. The worst plan is that seen in some of the northern dales where the manure is thrown out of a hole in the wall and left exposed to weather, with the result that streams of black liquid flow away. A much better plan is to cart the manure to a dungstead where it can be stored in a compact heap.

Liquid manure. The liquid manure draining from the cow sheds is rich in nitrogen and potash, and where practicable should be run into a tank and applied when convenient to the land. It may go on to grassland at almost any time, and to arable land after the autumn and before the middle or end of May.

Table XXIX. *Composition of liquid manure*

	Percentage composition			1000 gal. contain in lb.		
	Nitrogen (N)	Phosphate (P_2O_5)	Potash (K_2O)	Nitrogen (N)	Phosphate (P_2O_5)	Potash (K_2O)
English samples:						
Kent	0·18	0·017	0·40	18·2	1·7	40·1
Surrey	0·23	0·03	—	23·0	—	—
Scotch samples*	0·20	0·03	0·46	20·5	3·0	46·0

* Hendrick, *N. Scotland Coll. Agric. Bull.* no. 19, 1915; *J. Irish Dep. of Agric.* **13**, 251.

Storage and use of liquid manure are, however, frequently impracticable, and the dairy regulations usually require it to be removed forthwith.

The effect of cake feeding. Cake-fed dung, as shown on p. 233, is richer in nitrogen than dung produced on hay and roots only, and is even better than the figures indicate because the extra nitrogen is largely in the form of ammonia and amides produced from the liquid excreta. Further, cake-fed dung may rot down better in the heap than ordinary dung.

The difference in crop-producing value is shown in Table XXXI. It was very marked in the first year but not in the second or later years: the advantage apparently lies in the ammonia present and this persists for one year only.

Method of applying the dung. Directly farmyard manure is put on the land it begins to lose ammonia and should be ploughed in. At Rothamsted a delay of three weeks reduced the yield of sugar beet by 14 cwt. per acre, as compared with farmyard manure ploughed straight in.[1]

Application in the bouts was much more effective for

[1] Experiments in Germany by S. E. Scammell (*J. Min. Agric.* March 1936, p. 1226) also show the great loss resulting from delay in ploughing in.

Treatment	Relative yield of swedes
Full-dose dung ploughed in immediately after spreading	100
Full-dose dung ploughed in 1 day after spreading	94
Full-dose dung ploughed in 4 days after spreading	86
Half-dose dung ploughed in immediately after spreading	88
No dung	76

Some good Danish experiments afford further evidence.

potatoes than broadcasting the dung and ploughing it in, as shown in Table XXX.

Table XXX. *Yield of potatoes, tons per acre, obtained when dung was (a) ploughed in, (b) put into the bouts*

	No dung	Ploughed in	In the bouts	Advantage for application in bouts
1935	5·24	7·15	8·06	+0·91
1936	5·21	6·45	8·33	+1·88
1937	6·16	7·64	9·60	+1·96

It is more difficult to determine whether the manure is better applied in autumn or in spring. There are not many experiments and the recorded differences are not great: where there is an advantage it lies with the spring dressing.[1]

How long do the effects of farmyard manure last? Table XXXI shows that a dressing of dung given once in four

Table XXXI. *Comparisons of effects of cake-fed and ordinary dung. Little Hoos field. Produce per acre*

	Wheat bushels	Barley bushels	Roots tons	Clover hay cwt.
Ordinary dung				
Control, no dung	22·1	21·2	10·2	49·2
Additional yield for dung	8·2	16·9	3·4	20·6
In 2nd year	4·8	14·1	1·6	16·5
In 3rd year	4·4	11·4	1·6	15·2
In 4th year	3·9	8·2	0	12·0
Cake fed dung				
Control, no dung	22·1	21·2	10·2	49·2
Additional yield for dung	12·2	20·9	4·4	22·0
In 2nd year	8·0	16·4	2·8	20·0
In 3rd year	4·0	9·0	1·6	19·2
In 4th year	4·1	10·9	−0·4	15·6

[1] *Rothamsted Report*, 1937, p. 32. Some old experiments by Berry in the west of Scotland also showed an advantage of spring applications (*West Scotland Agric. Bull.* no. 65, 1914).

years on a rotation of crops produces a marked effect in its first year, and a less but quite distinct effect in the second, third, and fourth years. These residual effects are very important and show that the return from the farmyard manure cannot be estimated from 1 year's crop. These results probably represent those which a farmer might expect to obtain excepting that no clover was included in the rotation. Other experiments indicate that clover also benefits from the residues of farmyard manure and itself increases the productiveness of the soil.

Table XXXII. *Residual effects of dung on the succeeding crop (cwt. per acre)*

Dung applied to		Amount of dung tons	Succeeding crop	Mean yield	Increase for residues of dung
Potatoes	1916	10	Wheat grain	11·9	+2·4
Potatoes	1920	15	Wheat grain	17·8	+3·6
Potatoes	1936	15	Spring oats grain	20·2	+2·7
Kale	1932	15	Barley total produce	95·1	+12·2
Kale	1936	15	Barley grain	12·0	+2·2
Barley	1921	14	Clover 1921 green weight	9·2	+6·7
			Clover 1922 hay	45·5	+8·2
			Clover 1923 hay	13·0	+2·3

The residues of the dung have had about the same effect as a dressing of 1–1½ cwt. sulphate of ammonia, i.e. 23–35 lb. nitrogen.

COMPARISON OF FARMYARD MANURE WITH INORGANIC FERTILISERS

Comparison of farmyard manure with inorganic fertilisers is difficult because of the wide difference in quantity of plant food supplied. A 10 ton dressing of farmyard manure, which is rather on the small side, contains about as much nitrogen as $7\frac{3}{4}$ cwt. nitrate of soda or $5\frac{3}{4}$ cwt. sulphate of ammonia. The comparison was made on a basis of equal nutrients in Hansen's experiments on a light loam and on a sand at Askor (S. Jutland), and the farmyard manure almost always gave poorer results[1] than the mixture of nitrate of soda, superphosphate, and kainit, containing the same amounts of nitrogen, potash and phosphates. Of course this meant that the farmyard manure was given in unusually small dressings and so could not exert its full effect. In the Rothamsted and Woburn experiments dressings of 14 tons farmyard manure per acre (200 lb. nitrogen) are used, and the difficulty of comparison is obviated by having several plots receiving sulphate of ammonia or nitrate of soda, the dressings ranging up to 4 or more cwt. per acre. One of these plots usually gives results above, and another below, those given by farmyard manure, so that equivalent values can be estimated. In an experiment with kale at Woburn (Table XXXIII)

Table XXXIII. *Yield of kale* (*tons per acre*)

	Woburn 1932 Sulphate of ammonia				Woburn 1936 Sulphate of ammonia	
	None	0·2 cwt. N	0·4 cwt. N	0·8 cwt. N	0·4 cwt. N	0·8 cwt. N
No dung	13·29	17·76	19·67	24·36	10·14	13·67
Dung	19·19	21·24	23·67	28·74	13·14	15·49
Standard errors	±0·713				±0·357	

15 tons of dung containing 170 lb. nitrogen gave in three separate tests about the same yield as 39 lb. nitrogen

[1] Fr. Hansen and J. Hansen, *Tidsskrift for Landbrugets Planteavl*, 1913, **20**, 345.

as sulphate of ammonia which shows that 100 lb. of nitrogen in dung has in the first year a similar action to that of 22 lb. of nitrogen in sulphate of ammonia. The experiment has been repeated on other crops and the results vary with soil, crop, season, and size of dressing, but it usually appears that in the first year and in conditions comparable with those of ordinary practice:

100 parts nitrogen in farmyard manure = 20 to 30 parts nitrogen in sulphate of ammonia.[1]

The residual effects are more difficult to estimate because of the great variations in the results. A reasonable estimate is that when these are added on, 100 parts of farmyard manure nitrogen would be equal to about 35–55 parts of nitrogen in sulphate of ammonia or to 30–50 parts of nitrate of soda. These are also the results obtained on the long-continued experiments at Rothamsted, Woburn and Saxmundham, where the farmyard manure residues accumulate, but where, however, losses and secondary effects are probably more marked than in practice.

Field experiments have shown that the crop does not recover the whole of the nitrogen from nitrate of soda or sulphate of ammonia but in good conditions only about half to two-thirds. For farmyard manure the recovery is lower: only about one-fifth to one-third. On the long-continued dunged plots at Rothamsted and Woburn about one-third of the nitrogen remains in the soil and one-third is lost: the loss is about the same as for sulphate of ammonia which leaves no nitrogen in the soil.

[1] Data are given in *Rothamsted Report*, 1937, p. 31; *Fifty Years of Field Experiments at the Woburn Experimental Station*, pp. 149 and 228; and E. J. Russell and D. J. Watson; *Emp. J. Exp. Agric.* 1938, **6**, 268–92.

Little is known about the nitrogen which remains in the soil, but there is some evidence that some of it can slowly become available to the crop.

One of the barley plots at Rothamsted received farmyard manure annually for 20 years from 1852 to 1871, but nothing since. Alongside is a plot that has received farmyard manure annually from 1852 to the present

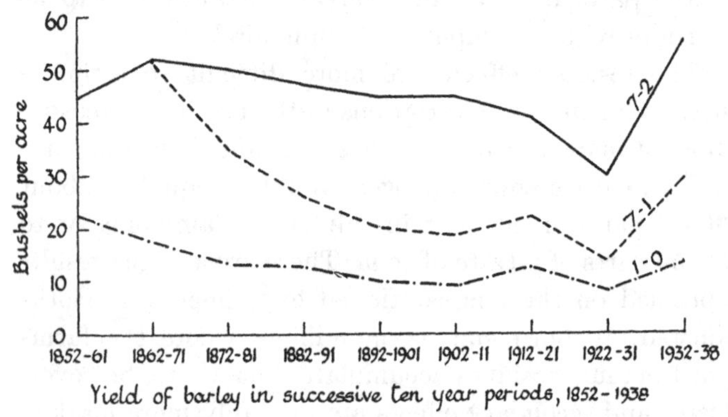

Fig. 51. Yield of barley in successive ten year periods, 1852–1938.

time, and another that has received nothing during the whole period. The effect of the farmyard manure went on increasing during the first 13 years; it then increased no more, but kept at its level. On the plot on which in 1872 the farmyard manure was discontinued the yield has gradually fallen, but even now it is still well above the level of the unmanured plot (Fig. 51).[1]

Farmyard manure contains rather more potash than

[1] See *Rothamsted Report*, 1937, p. 30, for a similar persistence of effect of farmyard manure.

nitrogen, and much of the potash is in an easily available form. Its phosphate content, however, is low (Table XXX), though some at any rate is easily available. If we assume full value for the potash and phosphate, and half value for the nitrogen, then a dressing of 10 tons farmyard manure is in its current and residual

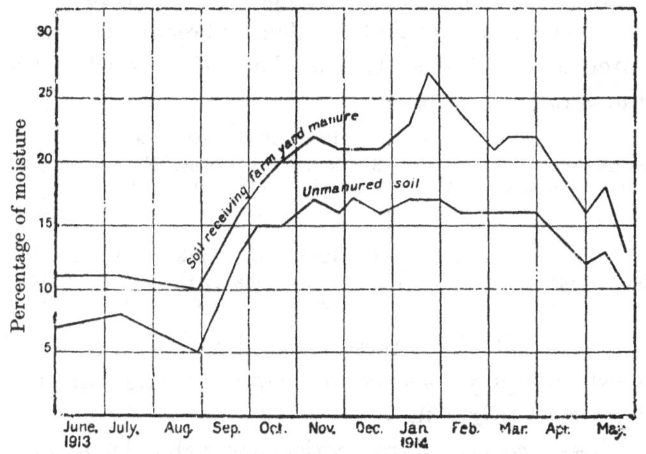

Fig. 52. Curves showing effect of farmyard manure on water content of soils. (Broadbalk field, Rothamsted.)

effects approximately equivalent to a dressing of 2–3 cwt. sulphate of ammonia, 3 cwt. muriate of potash and 4 cwt. superphosphate per acre.

The value of farmyard manure, however, cannot be completely assessed by the amount of plant food it supplies. It has a marked effect on the soil, improving both the tilth and the power of holding moisture which the plant can afterwards take up.

Fig. 52 gives curves showing the percentage of water in the dunged plot and in the adjacent unmanured plot of the Broadbalk wheat field during 1913–14: it will be seen that the former is invariably the moister even in the very dry June. Indeed, so great is the water-holding capacity of the soil that the rain water rarely gets down to the drains in sufficient quantity to cause them to run. The adjacent unmanured soil, on the other hand, is easily permeable to water; it becomes wet throughout its depth soon after rain has fallen, and readily transmits water to the drains. The numbers of days when the drains ran are as follows:

	1903	1904	1905	1906	1907	1908
Dunged plot	2	3	None	1	None	None
Unmanured plot	27	20	11	14	10	9

	1909	1910	1911	1912	1913	1914	Average of 12 years
Dunged plot	1	None	None	None	None	2	0·7
Unmanured plot	10	20	20	32	39	20	19·5

On mangold fields these physical effects on tilth and moisture supply considerably help the young plant in a dry season (Fig. 53).

These properties are particularly valuable on light soils in the drier parts of the country: e.g. in Norfolk, where in old days an efficient system of bullock feeding for dung production was in vogue. On heavier lands farmyard manure is not necessary for wheat or barley: most of the Broadbalk wheat field at Rothamsted, which has grown wheat every year since 1843, has had no dung since 1839 and yet it gives good wheat crops. But for sugar beet and potatoes, where more cultivation has to be done, farmyard manure appears to be more necessary.

Farmyard manure increases the exchangeable calcium in the soil, in virtue of the calcium present in the straw and the animal excretions: it also tends to keep the soil reaction more nearly neutral.

In consequence of these various effects on the soil it is not possible to set up a close comparison of farmyard manure and artificial fertilisers. In practice they should be used together, not regarded as rivals. Farmyard manure has generally in the Rothamsted experiments

Fig. 53. Barnfield mangolds in a dry season. The part of the field to the right of *A* has received farmyard manure, and the plants are growing well; the part to the left has not.

enhanced the effect of sulphate of ammonia on potatoes, though it has not done this for potassic and phosphatic fertilisers. Indeed, it rather depressed the effectiveness of the potassic fertilisers,[1] but this action is not marked and is more than counterbalanced by the benefits in other directions.

The best place in the rotation for the dung is the root

[1] *Rothamsted Report*, 1937, p. 34.

or potato crop, and, if it is equally convenient, application in spring is better than in autumn. The dung should go in the bouts and it should be covered directly it is applied; serious loss takes place if it is left exposed to air.

If dung can be spared some can go as top dressing in autumn on the young seeds and on the meadowland.

Unexhausted values. Farmyard manure is in rather a different category from the artificial nitrogenous fertiliser in that its effects are not confined to the season of application but persist over several years. So long as a farmer continues in possession of the land he may hope to gain the benefit, but if he gives it up before the effects have come to an end he is entitled to compensation for the unexhausted value of the manure. The first tables for the guidance of valuers were drawn up by Lawes and Gilbert in 1870; they were reissued in 1914[1] and periodically revised by Voelcker and Hall, who recommended that (*a*) compensation should be payable in respect of half the nitrogen and three-quarters of the potash and phosphoric acid contained in the food, it being supposed that the remainder is lost: this amount to be paid in full where the manure has been applied to the land but no crop grown; only one-half the above amount is to be paid after the growth of one crop. Values calculated on this basis are given in Table XXXIV. The Scottish valuers also adopt these proposals except that they allow compensation only for 40 per cent, and not 50 per cent, of the nitrogen in the food stuffs. The actual values need recalculating whenever the prices of fertilisers change: the values used in the Table are 10*s*. per unit for nitrogen, 3*s*. per unit for

[1] *J. Roy. Agric. Soc.* 1914, **74**, 104.

phosphoric acid, and 3*s*. 6*d*. per unit for potash. The change to any other price level is easily made. In the tables periodically issued in the *Journal of the Ministry of Agriculture* the nitrogen is valued at 7*s*. per unit, the phosphoric acid at 2*s*. and the potash at 3*s*.

There is no good experimental basis for any of these values, and at Woburn it was impossible to obtain anything like the benefits expected. Grazing experiments carried out under the Royal Agricultural Society at Shoby in Leicestershire also failed to show the benefits anticipated, and this question is now under examination at Rothamsted, as part of the Society's Research Scheme.

COMPOSTS

A compost is a heap of mixed vegetable and animal matter put up so that it can decompose and form useful organic manure. The art of making composts was well known to agriculturalists during the fifties and sixties of the last century, and farmers and gardeners made great use of them before artificial fertilisers were available. Indeed, in the *Gardeners Chronicle* for 1845, Mr Errington describes no fewer than twenty different composts for garden purposes, and explains their uses. A full account is given by Mr Hannam of Kirk Deighton, Wetherby, in Morton's *Cyclopaedia of Agriculture* in 1855.

The subject was put on a sound basis by E. H. Richards and H. B. Hutchinson who in 1921[1] pointed out that the decomposition was brought about by micro-organisms and could proceed satisfactorily only if they

[1] *J. Min. Agric.* 1921, **28**, 389 and 482.

were given their essential conditions for activity: adequate supplies of food, water, and air, and a neutral or nearly neutral reaction. The materials usually available contained inadequate amounts of nitrogen, and the heap tended to go acid. All these defects could easily be remedied. Analysis of the material showed whether more nitrogen should be added and if so, how much: it could be given in any available form, e.g. as sulphate of ammonia, cyanamide, etc., along with calcium carbonate and a little phosphate if necessary. This mixture was sprinkled on the successive layers as the heap was made, water was then added, and decomposition proceeded satisfactorily. The process was worked out in detail, and applied on the large scale. It is now much used by market gardeners for converting straw into manure.

Several variations of the method are also in use, the difference lying in the way the organisms are fed: vegetable or animal matter can be used, household wastes, etc., instead of a simple nitrogen compound.

The growing shortage of stable manure and farmyard manure has brought the subject of composting again into importance.

TOWN REFUSE

It is estimated that over 8 million tons a year of town refuse containing more than 1 million tons of vegetable and other matter of fertiliser value have to be disposed of annually and the problem is becoming of increasing importance. The present cost to the public is about £2½ million annually. In some places farmers have been willing to use town waste as manure, but large quantities have to be dumped or carried out to sea.

As it stands it is not very attractive to farmers. It contains vegetable and animal matter of fertiliser value mixed with a good deal of material of no value, even detrimental on the land, and its character has in some ways deteriorated in recent years. The proportion of paper seems to be increasing, and amounted to 14 per cent in the samples examined in 1935–6: this not only has no fertiliser value but it is detrimental in that its decomposition leads to an assimilation of nitrate by the micro-organisms concerned. Further, the number of tins had also increased to 46 per head per annum—and as the corresponding number in the United States was 94 there seems the probability of much further increase. These are mechanically objectionable. On the other hand, cinders have some advantages on heavy soils though not, so far as is known, on light soils. Cinders, however, tend to decrease. Some calcium carbonate is present, but the vegetable and animal matter is the chief ingredient of fertiliser value; its proportion varies according to the class of house and the season. It constituted on the average about 13 per cent in the samples examined; from the better class houses it was 16 per cent and from artisan houses 10 per cent; in a slum area it was as low as 3 per cent. It was greater in summer (up to 28 per cent in the better class houses) than in winter.[1]

It is, however, the other ingredients that make the material difficult to handle and these seem to be increasing in amount.

[1] See H. Edridge's *Report*, Institute Public Cleansing Annual Conference, June 1937, and E. J. Russell, *J. Roy. Agric. Soc.* 1937, **98**, 402.

Recently methods have been adopted for removing usable material such as cinders, glass, metals, and also paper and cardboard: this improves the residue as fertiliser. The percentage of nitrogen in ordinary waste is usually 0·5 or less: in a sorted waste it was 0·7: i.e. about 1 per cent on the dry matter since the waste contained 25 per cent of moisture.

In the year of application a sorted and treated sample examined at Rothamsted in 1938, which may be typical of the samples of the future, proved better than farmyard manure supplying equal amount of nitrogen and not quite as good as sulphate of ammonia or rape cake supplying half the amount of nitrogen.[1] No comparison of residual effects has been made, nor is it known what physical effects may be produced on the soil. The problem of utilising this material is very urgent.

[1] This is not an exception to the rule given on p. 209: the percentage of nitrogen in the organic matter was 2·9, which is well above the limit of 1·5 per cent stated in the rule.

CHAPTER XIV

CHALK, LIMESTONE AND LIME

CALCIUM carbonate is an important constituent of soils because it constitues the reserve from which supplies of exchangeable calcium are maintained and it ensures a neutral reaction such as is desirable for crop growth (p. 120). Soils vary enormously, however, in the quantities they contain: the variation may be from 20 per cent or more in the chalk soils down to nothing in very acid soils. Non-calcareous soils may contain about 0·25–1 per cent, and these need constant watchfulness as the lime is very liable to loss. Under an annual rainfall of about 30 in. the soil at Rothamsted has been estimated to lose calcium carbonate at an average rate of about 800 lb. per acre a year. Under higher rainfall the losses are greater. Sulphate of ammonia also increases the loss by an amount nearly equal to its own weight: other fertilisers do not (p. 170). This loss of lime has but little effect till the soil reaction falls below neutrality: the acidity then setting in is unfavourable to many crops and natural plants and a typical weed flora of acid tolerant plants appears (p. 121). Certain disease organisms also flourish, such as *Plasmodiophora* which causes finger-and-toe in turnips and other plants of the Brassica tribe.

The remedy consists in adding lime in some suitable form. The need has been well known for very many years. Some of the first agricultural colonists in England, the Belgic peoples who came here before the Romans, dug wells to get out the underlying chalk and spread it on the surface as manure. The method continued in use till about 1914. Large dressings were given, even 50 or 80 tons per acre, and they had the effect of improving the texture of the soil, and notably facilitating cultivation and drainage.

Since then these large dressings have been uneconomic, and for some years chalking and liming fell into abeyance. With the Government Land Fertility Scheme, however, the use of both lime and basic slag has greatly extended.

The substances now used are:

Burnt lime (calcium oxide, CaO),

Hydrated lime [$Ca(OH)_2$],

Carbonate of lime ($CaCO_3$) in various forms: chiefly chalk and limestone.

Calcium oxide is the active agent in improving the soil, and burnt lime is therefore the most concentrated of the above forms, 1 ton being equivalent to nearly 36 cwt. of calcium carbonate. Apart from this, however, it has no special advantage over the carbonate; indeed, in the older experiments it came out inferior. Unfortunately no good modern experiments have been made.

It must be ground fairly finely in order to allow for distribution, but it is then apt to be disagreeable for the workers. Further, when lime is put on to grassland there is some tendency for it to cake after rain. It probably works better into arable land.

Low-grade burnt limes are periodically available and although the percentage of CaO may be low the other constituents, especially the carbonate and some of the silicates, have distinct neutralising value. In general they should be bought on analysis.

Hydrated lime is easier to handle but it is less concentrated, 100 parts containing 76 parts CaO, the rest being combined water of no agricultural value.

Calcium carbonate is available in several forms. Chalk is the softest form: it disintegrates and dissolves in the soil so that it need not be finely ground. Limestone is much harder, and should be ground, not used as lumps. Finally, like chalk, it dissolves in the soil moisture with the help of the carbonic acid and then enters into reaction with the clay, the humus, and other soil constituents.

Sugar beet factories have for disposal a waste lime which contains some 30–40 per cent of calcium carbonate and is often obtainable at low cost on the farm.

Magnesian limestone is in some districts less popular than ordinary limestone but there is no evidence that it is in any way inferior. In the old days the magnesium was supposed to be harmful: actually it is not, although if a heap of burnt magnesian lime were left standing in a field the alkalinity of the magnesia might destroy the underlying vegetation. But as used nowadays its action is, so far as is known, as good as that of ordinary burnt lime.

It is relatively easy to find out how much lime has to be added to a soil in order to make it neutral, or sufficiently nearly neutral for crops to grow well. The "lime requirement" is determined by analytical methods, and the degree of acidity on the *p*H scale by a colorimetric or more exact electrometric test. These tests should always be made so as to ensure using the right amount of lime.

The amount of lime added must always be sufficient, and, as shown above, there are some advantages in having an abundant supply of the carbonate in the soil.

But there are also disadvantages. On some soils it increases the liability to deficiency diseases, e.g. boron and manganese; and on all soils the rate of loss of lime is greater when an excess is present than when there is only sufficient.

Ground limestone or lime should be applied in autumn or early spring and may with advantage be used for sugar beet, the clover crop or swedes, but not for potatoes owing to the risk of scab.

Ploughed out grassland should always be tested to see if it needs lime, and, if so, how much.

Table XXXIV *Voelcker and Hall's table. (Revised by Dr Voelcker, June,* 1937.*) Showing the composition, manurial and compensation values of feeding stuffs*, available for download from www. cambridge.org/9781107654341

CHAPTER XV

MANURIAL SCHEMES FOR CROPS

FERTILISERS and manures are used for two purposes: to make good any serious deficiencies in the soil; and to raise the level of crop production in soils where no marked deficiencies exist. In this chapter we shall deal only with this second purpose.

ARABLE CROPS

The manurial scheme must be worked out for the rotation and it should centre round the root and potato crops, as these are usually the most responsive and also they are the most capable of giving the highest yields of dry matter per acre. The basis should be farmyard manure at the rate of about 15 tons per acre applied in spring or autumn but with a preference for spring; ploughed under as soon as the general work of the farm permits. The following recipes have given good results but should be modified to suit the conditions of the farm.

Potatoes. This is one of the most important crops on the farm and usually responds to generous manurial treatment. Numerous experiments made by the Rothamsted staff in various parts of the country have shown that potatoes respond to potash on light soils but less on heavy soils or where farmyard manure is given; they respond well to phosphate on practically all soils. The effect of fertilisers on the proportion of ware was also marked; nitrogenous fertilisers increased the percentage of ware on heavy soils but potash and phosphate did not. On light soils, on the other hand, nitrogenous fertilisers rather reduced the percentage of ware but phosphate increased it. Potash had no effect when dung was given but was markedly beneficial in absence of dung. The results are given in Table XXXV.

Table XXXV. *Effects of fertilisers on yield of potatoes (tons per acre) and on percentage of ware on light and heavy fenland soils*, 1935

	Light Fen Thorney (with dung)		Light Fen Mepal (no dung)		Heavy Fen March (no dung)		Heavy Fen Little Downham (no dung)
	Yield tons	Ware %	Yield tons	Ware %	Yield tons	Ware %	Yield tons
Mean	8·40	85·8	9·49	61·7	6·80	82·4	5·02
Increase for:							
3 cwt. sulphate of ammonia	+0·25	−2·1	+1·31	−6·2	+2·90	+5·4	+3·03
9 cwt. superphosphate	+0·95	+4·3	+1·08	+3·3	+1·17	+0·7	+1·90
3 cwt. sulphate of potash	+0·99	−2·0	+4·31	+22·2	−0·20	−1·1	+0·30

The experiments bring out the important property of fertilisers to which reference has already been made: that one fertiliser not infrequently reinforces the effects of another. On the heavy soil the effect of the double dose of nitrogen is increased by superphosphate but not by potash. On the light soil, on the other hand, it is potash rather than phosphate which increases the effectiveness of the nitrogen (Table XXXVI).

The following manurial scheme is suitable for trial:

Ten to fifteen tons per acre of farmyard manure applied in the bouts if practicable. A mixture of 2–3 cwt. sulphate

Table XXXVI. *Effect of one fertiliser in reinforcing the action of another: nitrogenous fertiliser reinforced by phosphate on heavy soil and by potash on light soil*

Heavy soil, Little Downham, Ely, 1935.

Yield tons per acre ±0·336.[1]

Sulphate of ammonia cwt.	Superphosphate cwt. per acre 0	4½	9	Sulphate of potash cwt. per acre 0	1½	3	Mean ±0·194
0	3·09	3·83	3·57	3·30	3·70	3·49	3·50
1½	3·85	5·30	5·94	4·50	5·04	5·54	5·03
3	4·79	6·89	7·92	6·40	7·14	6·06	6·53
Mean ±0·194	3·91	5·34	5·81	4·73	5·29	5·03	5·02

Light soil.

Yield tons per acre ±0·718

Sulphate of ammonia cwt.	Superphosphate cwt. per acre 0	4½	9	Sulphate of potash cwt. per acre 0	1½	3	Mean ±0·414
0	8·01	8·88	9·55	6·96	9·14	10·34	8·81
1½	8·99	10·37	9·27	7·57	9·09	11·97	9·54
3	9·73	9·47	11·14	6·96	11·29	12·10	10·12
Mean ±0·414	8·91	9·57	9·99	7·16	9·84	11·47	9·49

[1] This figure is the "standard error", see Note on p. 271.

of ammonia, 2–3 cwt. sulphate or muriate of potash, and 4 cwt. superphosphate, at the time of planting.

Where 10 tons or more per acre of potatoes are expected, the quantities of the fertilisers can be increased by 50 per cent.

Where, however, the yield does not normally exceed 7 or 8 tons per acre, and cannot be raised above this level by the use of fertilisers, the dressing can be reduced to 2 cwt. sulphate of ammonia, 1 cwt. sulphate or muriate of potash, and 3 cwt. of superphosphate.

There is little to choose between the two potassic fertilisers in regard to yield, but the sulphate gives the better quality and should be used wherever this is an important consideration. The muriate, however, serves quite well where the potatoes are sold straight off the ground for ordinary household purposes. Potash salt is less suitable.

Top dressings are not usually necessary or desirable for potatoes.

Mangolds. 10 to 15 tons per acre farmyard manure. At seeding time 2 cwt. sulphate of ammonia, 3 cwt. superphosphate and 3 cwt. 30 per cent potash salt: on some soils 2 cwt. of agricultural salt. At the time of singling 1½ cwt. nitrate of soda can be given as a top dressing. This seems a generous manuring, but the mangold is a responsive crop. Where, however, the yields do not justify so much expenditure the dressings can be reduced.

Sugar beet. It is more difficult to give general recommendations for sugar beet as the experiments are still in progress. If the soil is acid lime should be added since sugar-beet is very sensitive to acidity. Farmyard manure should be the basis of the manuring: nitrate of soda is the best of the nitrogenous manures; salt is beneficial; and of the potassic fertilisers 30 per cent potash salt appears to be better than the sulphate or the muriate.

The tops are usually increased by the nitrogenous manure, and where they are fed they may even repay the cost of the manure.

The following is suggested as a basis for trial: 10 tons farmyard manure ploughed under in autumn, 2 to 3 cwt. nitrate of soda, 3 cwt. superphosphate, 3 cwt. potash salt per acre applied at or before sowing, the nitrate of soda being put on separately. Trial should be made of the effect of 2 cwt. salt per acre instead of the potash salt.

Swedes. When high yields can be obtained, e.g. 35 or more tons per acre, farmyard manure should form the basis of the manuring. At the time of sowing: 1–2 cwt. sulphate of ammonia, 5 cwt. superphosphate, and 1 cwt. muriate of potash can be given.

In the central and southern parts of England where the yields are commonly less than 25 tons, there is no advantage in using both farmyard manure and artificials—either can be used, whichever is convenient, but not both. If artificials are chosen, a suitable dressing is 1 cwt. sulphate of ammonia and 4 cwt. superphosphate. Where the land is liable to finger-and-toe, it is necessary to apply lime before using superphosphate; alternatively, high-soluble basic slag can be used.

Kale. Farmyard manure as for mangolds and potatoes: 2½ cwt. superphosphate, 2 cwt. 30 per cent potash salt and 1 cwt. sulphate of ammonia as basal manure, and dressings of nitrochalk when the crop is safely up and out of danger of harm by the flea beetle. The top dressings can vary from 1 to 3 cwt. per acre or sometimes more; this can be settled only by trial.

CEREALS

After the roots or potatoes a cereal crop would usually be taken, and generally sufficient manurial residues are in the soil to carry the crop without further addition of fertiliser.

If, however, the winter has been very wet so much nitrate may have been washed out from the soil as to make a top dressing desirable for winter oats or wheat, and in that case 1–1½ cwt. sulphate of ammonia, nitrate of soda or nitrochalk can be given. Late applications give as much grain as early ones, but usually they give less straw.

Spring-sown barley would not normally follow potatoes but it can follow sugar beet. After a dry winter little if any fertiliser would be needed, but after a wet winter 1 cwt. sulphate of ammonia should be given at the time of seeding.

Where barley follows a cereal crop it should receive 1 cwt. sulphate of ammonia and, on chalk soils, 1 to 2 cwt. of 30 per cent potash salt.

Clover or a seeds mixture is usually sown in the first or second cereal crop and usually receives no fertiliser. On soils liable to acidity this may be a good place to apply lime or calcium carbonate: young clover and barley are very sensitive to acidity.

After the seeds ley another corn crop is usually taken and either receives no manure or a dressing of nitrogenous manure.

OTHER ROTATIONS

The rotation indicated above: roots, corn, clover or seeds, corn, is very common in Great Britain, but many others are in use. Here are a few examples:

(I) On the very fertile soils of the Holland division of Lincolnshire:

(1) Main crop potatoes, well manured[1]
(2) Mustard or wheat
(3) Wheat or oats
(4) Clover or peas

[1] Foundation of dung, followed by artificials. The bigger farmers often use 15 to 17½ cwt. per acre of a mixture of 3 parts sulphate of ammonia, 4 parts super, 2 parts sulphate or muriate of potash, and 1 part drier.

or:

(1) Main crop potatoes, well manured
(2) Sugar beet
(3) Cereals
(4) Clover, peas, or beans. In this case the sugar beet often needs no fertiliser, though in other cases 5 to 7 cwt. per acre of a mixed "compound" are given. The cereals do not require fertiliser.

In both rotations the potatoes follow a leguminous crop.

(II) Cambridge University Farm:

(*a*) *Light land*

1st year	Potatoes	12 loads farmyard manure 3 cwt. sulphate of ammonia 3 cwt. superphosphate 2 cwt. sulphate of potash
2nd year	Wheat	No manure
3rd year	Sugar beet	12 loads farmyard manure 2 cwt. sulphate of ammonia 3 cwt. superphosphate $1\frac{1}{4}$ cwt. muriate of potash
4th year	Barley	No manure
5th year	Kale, etc.	2 cwt. sulphate of ammonia 3 cwt. superphosphate 1 cwt. muriate of potash
6th year	Barley	No manure

(*b*) *Heavy land*

1st year	Single-cut seeds followed by bastard fallow	No manure
2nd year	Wheat	No manure
3rd year	Winter oats	2 cwt. sulphate of ammonia 2 cwt. superphosphate
4th year	Silage (followed by bastard fallow) or beans	12 loads farmyard manure 6 cwt. superphosphate
5th year	Wheat	No manure
6th year	Wheat	12 loads farmyard manure 1 cwt. nitrochalk 2 cwt. superphosphate

(III) South-Eastern Agricultural College, Wye:

(*a*) *Light land*

Three years temporary ley. On breaking ley:

1st year	Peas	2 cwt. superphosphate 35 per cent 1 cwt. steamed bone flour 1 cwt. muriate of potash 1 cwt. sulphate of ammonia
2nd year	Wheat	1 cwt. sulphate of ammonia
3rd year	Barley	Same mixture as for peas
4th year	Barley	6 cwt. basic slag 40 per cent 1 cwt. muriate of potash 1 cwt. sulphate of ammonia 1 cwt. steamed bone flour

Seeds are sown in this barley and left as temporary grass for three years.

The mixture consists of:

4 lb. wild white clover
12–16 lb. perennial rye grass (grazing strain)
2 lb. rough stalked meadow grass

(*b*) *Stronger land*

As for light land except that

3rd year Barley is replaced by wheat which, however, also receives the same manure as for peas.

The peas may be replaced by potatoes, mangolds, kale, wheat or oats. If potatoes are grown 12 to 16 loads of farmyard manure are given, the quantities of artificials are doubled, and sulphate of potash is used in place of muriate of potash. On breaking the leys skin coulters are used to bury the turf and furrow pressing is also carried out. Much importance is attached to this.

Permanent pastures are broken up in early spring and cropped as above, but after the peas are harvested (July) the land is well worked, being ploughed at least twice, to reduce weeds, wireworms and other pests.

GRASSLAND

The general rule is that nitrogenous fertilisers give bulk, potash and phosphate give quality, while on acid soils lime is necessary to maintain both yield and quality.

Permanent hayland. Satisfactory results have been obtained with farmyard manure every 4 or 5 years and artificials in the intervening years: 4 cwt. of high-soluble slag and 3 cwt. of 30 per cent potash salt once in 4 years, and 1 cwt. nitrate of soda or sulphate of ammonia each spring except in the season when farmyard manure has been given. Periodically a dressing of ground limestone or chalk is necessary; the need for this can be judged by the character of the herbage.

Temporary hayland. Sulphate of ammonia applied to the barley crop is apt to depress the yield of the seeds mixture, while muriate of potash or potash salt may make the clover grow too vigorously and give trouble at harvest. The difficulty can be overcome by the application of potassic fertilizer and, if necessary, basic slag, after the barley has been cut. For a one-year ley it is unnecessary to add nitrogenous fertilisers, but for longer leys, 1 cwt. of sulphate of ammonia or nitrate of soda in the spring of the second and succeeding years may increase bulk without spoiling the herbage.

New grass. A good dressing of basic slag and if necessary potash salt should be given at the time of sowing. No nitrogenous manure is needed; sulphate of ammonia should not be given until the clover is well established. A top dressing of rotted dung is very effective.

Grazing land. The standard method of improving grazing land is to treat it with basic slag at the rate of 5–10 cwt. per acre. The slag should be of high solubility—80 per cent or more in 2 per cent citric acid.

Grazing, however, is at least as important as manuring, and, unless it is properly managed, pasture land may derive no benefit from manuring. It is imperative that the herbage should be kept down as closely as practicable, and for this reason mixed grazing is usually better than grazing by a single type of animal: there are, however, certain exceptions, such as the sheep-pastures of Romney Marsh.

VEGETABLE CROPS

Experimental work on vegetables is more difficult than that on farm crops, and only little has been attempted till recently. The general soil conditions are of particular importance and the basis of the manuring must be organic matter applied sufficiently often and in sufficient quantities to build up good crumb structure and ensure adequate water-holding capacity. By the time this is done the soil is usually very rich in plant food and fertilisers are needed only for special purposes. Some of the most successful growers use very little; Mr F. G. Secrett states that he spent annually on 200 acres of vegetables over £4600 on horse manure and £526 on organic manure, but only £255 on ordinary artificial fertilisers.[1]

Many growers, however, are unable to obtain sufficient

[1] *Journal, Royal Society of Arts*, 1935, **84**, 126. In agreement with this the Ministry of Agriculture "Grow More" leaflet recommends for a 10 rod allotment a dressing of 20 lb. sulphate of potash and 30 lb. superphosphate in spring, followed by 15 lb. sulphate of ammonia to the growing crops as required.

organic manure and must therefore maintain the plant food supply by adding the necessary fertilisers. The method differs in different parts of the country, but the problem is always complicated by the fact that total yield is only part of the growers' requirements; earliness, freedom from disease, firmness of produce like sprouts, cabbage, broccoli, and other properties summed up under the name "quality" are vitally important.

In the market-garden regions of Bedfordshire, Mr J. W. Dallas recommends applications of sufficient potash and phosphate, and, if necessary, lime, to ensure that the soil remains in good heart, but it seems immaterial to which crops these are added. Nitrogen, however, is in a different position: here the time factor is important; nitrogen must be given to the appropriate crop at the right time. Season, bad soil conditions, and previous cropping and manuring play so important a part that no general receipts can be given; experiments are needed in each area and on the different soil types. Compound fertilisers are not widely applicable for market-garden crops: a mixture may suit one set of conditions but not another: the need of each crop must be considered in relation to all the circumstances in which it is grown, though most crops benefit from a nitrogenous fertiliser properly used.

In market-garden districts of Cornwall very heavy manuring is adopted for some of the crops, especially for the early potatoes of the Penzance region. The soil is light, and the climate very moist and warm.

A dressing of compost[1] (40 loads per acre) in February

[1] Called "pile" and made of farmyard manure, sea weed, town refuse, broccoli stumps, sand, etc.: an ancient method in Cornwall. H. W. Abbiss, *Potato Production in Cornwall*, 1937.

is followed by a dressing of 1 part sulphate of ammonia : 3 parts superphosphate, 1 part steamed bone flour and 1½ parts sulphate of potash at the rate of 10 cwt. per acre, or alternatively 1 ton per acre of fish manure or Peruvian guano; when the tops are above ground 10 cwt. per acre of nitrate of soda are given. Planting in the Mounts Bay area begins in the first week in February: the sets are put very closely, about 8 to 10 in. apart and the rows are at about the same distance; about 2 tons per acre of sprouted seed are used. The crop is ready for lifting during May and early June. The yields vary from about 7 tons per acre for the first diggings in early May to 10 or 12 tons per acre in early June. The cost of production is high: the rent of the land may be as much as £10 per acre, but the value of the crop often exceeds £100 per acre and the high manuring is justified.[1]

Broccoli usually follows. This seems to do better on the residues of farmyard manure than on the fresh material, so that no more is added. Potash is essential, however, to enable the plant to resist cold and fungus disease (see p. 192) and also to produce solid "curds" that will reach the market undamaged by the long journey. A dressing of 4 cwt. superphosphate or basic slag and 4 cwt. sulphate of potash per acre, or 5 cwt. coarse hoof, 5 cwt. bone meal, and 3 cwt. sulphate of potash, may be harrowed in before planting. 2 to 3 cwt. of fish guano may be given before the plants are "banked"—done to protect them against high wind. Later on, top dressings of liquid manure and of 1 to 2 cwt. per acre nitrochalk are given.[2]

[1] The earliest potatoes may sell for £30 per ton, and the early June liftings may realise £18.

[2] H. W. Abbiss, *Winter Cauliflower or Broccoli*. (Cornwall County Council.

This sequence of early potatoes and broccoli is adopted also on the loams in the district round the Wash, where, however, the harvests are later than in Cornwall. The mixture recommended by Mr J. C. Wallace for early potatoes consists of 2½ parts sulphate of ammonia, 4 parts superphosphate, 1½ parts muriate of potash and, if necessary, 1 part of drier (e.g. steamed bone flour) applied at the rate of 12 cwt. per acre, or rather more if a drier is used: this differs from the Cornish mixture in containing nearly twice as much nitrogen. Top dressings of nitrate of soda or sulphate of ammonia have not, however, proved useful and so are not given: this difference from Cornish practice may arise from the circumstance that the rainfall is much lower than in Cornwall.

Broccoli may follow, receiving 3 cwt. superphosphate and 1 cwt. muriate of potash at planting time and 2 cwt. nitrate of soda or nitrochalk in spring.

Alternatively, spring cabbage may be planted in October or November and given three dressings of nitrate of soda, each of 1 cwt. per acre, the first being in early February. Sulphate of ammonia gives somewhat the same yield, but the crop develops more slowly and it may suffer from scorching if the sulphate is applied carelessly.

Autumn cauliflower may follow without fertiliser. In other cases, however, cauliflower receives heavy dressings of farmyard manure and then a mixture of 2 cwt. sulphate of ammonia, 4 cwt. superphosphate, and 1½ cwt. muriate of potash per acre.

Celery is another well-manured crop: it usually has farmyard manure and then a top dressing of 2 to 3 cwt. sulphate of ammonia. Lettuce, on the other hand, is in the nature of a catch crop, but in the experiments it still responded to a dressing of 4 cwt. superphosphate and 2 cwt. nitrate of soda or nitrochalk per acre, giving more first-grade plants and earlier cutting. Potash proved ineffective.

Carrots are also a catch crop but receive no fertilisers.

At Studley College, Warwickshire, the soil is heavier and the crops later than in the Wash regions just described, and the manuring differs in that less nitrogen is used and no potash except in the farmyard manure. Late potatoes receive well-rotted manure and 4 cwt. superphosphate per acre: earlies receive soot, bone flour, or stable manure according to previous crops. Cabbages follow: they receive basic slag and bone meal at planting, then in March a top dressing of sulphate of ammonia or soot. Lettuce receives a mixture of 3 parts bone flour and 1 of sulphate of ammonia at sowing time. Celery, as in Lincolnshire, receives stable manure, but soot at planting time in place of the top dressing of sulphate of ammonia. For onions, liberal quantities of farmyard manure and old soot are given, the former dug in in autumn and the latter at sowing time. Parsnips without manure may follow either of these.

The manuring of market-garden crops is under investigation both at Rothamsted and at the Horticultural Research Station at Cambridge.

The results so far obtained in the experiments made by the Rothamsted Staff show that the leafy crops like cabbage, sprouts, etc., respond well to manures, especially nitrogenous manures; that the organic manure used, dried poultry manure, is less effective than artificials,[1] the nitrogen having only about 65 per cent the

[1] See also A. H. Lewis, *Empire J. Exp. Agric.* 1938, **6**, 38–52.

value of that in sulphate of ammonia, but that on Brussels sprouts it sometimes had certain effects of earliness, and, on some soils, of persistence, not shown by the artificial fertilisers; and that some crops, notably carrots, onions, lettuce and strawberries, showed little response to artificial fertilisers.

Peas picked green and for canning: French and Runner beans. Nitrochalk and sulphate of ammonia gave good increases when applied at the rate of 2 cwt. nitrochalk per acre.

Brussels sprouts. The most marked effect is that of nitrogenous fertiliser and this varies considerably from season to season.

Cabbages and broccoli. Spring cabbages and broccoli both respond well to sulphate of ammonia or nitrate of soda, increasing both yield per acre and number of plants ready to cut by the beginning of December.

THE DRILLING OF FERTILISERS

The old method of using fertilisers was to broadcast them in a powdered form and then to harrow them into the soil. Later they were drilled, and this method was developed on the great wheat farms of North America and Australia where they are put in with the seed, the drill carrying the two boxes, one for the manure, the other for the seed. A still further improvement could be made if it were practicable to place the fertiliser alongside the seed as the plant then makes better use of it: drills have been made for this purpose but are not in common use.

In recent years fertilisers have been granulated, and this facilitates distribution and ensures a complete mixing of all the constituents.

APPENDICES

I. FIELD EXPERIMENTS[1]

While chemical analysis and pot cultures give useful information about the properties of fertilisers, the final test of their value on the farm must always be their behaviour in the field.

The old method of making field experiments was simple and attractive. A series of plots was laid out on as uniform an area as could be found; one of these was left as control, and the others were treated in various ways. The produce was then weighed, and it was assumed that the differences were due to the treatments.

Most of our knowledge of the effects of fertilisers on crops and soils was obtained in this way. The classical Rothamsted plots, and the later ones at Woburn and Cockle Park, are all of this design. The information obtained in good experiments was true in a general way, but it was lacking in precision; its weak point was that one could never say with certainty how much of the difference in yield between the treated and the control plot was due to the treatment, and how much to soil difference between the plots. In spite of this uncertainty, however, the method served very well for the early stages of agricultural science, and indeed it is still often used, and quite effectively, for demonstration purposes. The figures that it gives, however, cannot usually be accepted at their full value.

[1] In the presentation of this difficult subject I wish to acknowledge much help from Mr H. W. Gardner.

Advances in scientific knowledge have generally come from improvements in measurement, and in recent years investigators have endeavoured to work out new and better methods of field experiments.

The chief difficulty to be overcome has been the inequality of the soil: no field is absolutely uniform, however carefully it is chosen. Even the Broadbalk wheat field, the best known experimental field in the world, is somewhat uneven, and during the early years when the comparison was made there was a difference of nearly a bushel per acre between the unmanured plots at the two ends of the field. The differences due to treatments were all much larger than this and hence can be accepted in a general way, though the possibility of irregularity still remains.

It is not always possible, however, to find so uniform a field as this. Also the wheat crop has far more plants per unit area than fruit or vegetable crops, so that individual irregularities have more chance of being smoothed out.

The older method of overcoming these difficulties was to repeat some or all of the plots, and then try to make some allowance for differences in the soils or plants. This method proved useful in certain cases but not in others, and has been given up by most workers.

The modern method recognises the fact that no experiment, whether made in the laboratory or in the field, is ever free from errors due to soil irregularities and other causes, and experiments should therefore be designed that the amount of the error can be calculated, and the investigator may know what degree of significance attaches to his results. R. A. Fisher and his successors in the Statistical Department at Rothamsted have worked out ways of doing this which are now very widely adopted. The experiment is so arranged as to give a number of independent comparisons, some of which are between the treatments, some deal with soil variation, and the rest are between plots or groups of plots which have had the same treatment. Some repetition or replication of the treatments is necessary to allow error differences to be sorted out from treatment differences. In the case of field experiments the "replicates" are arranged in strips or "blocks" so that the differences between the areas constituting the blocks may affect all the treatments as equally as possible. In this way some of the major soil or position differences are allowed for.

The yields of the individual plots then give a number of independent comparisons which can be divided into three sets:

(1) between treatments;
(2) between blocks or strips;
(3) between groupings of plots which have been taken equally from all the blocks.

Set (3) gives a measure of the variation or experimental error which is neither due to treatment nor to block differences, and this measure is used as a test of the reliability of any differences between the treatments. If it is small, then even small treatment differences may be accepted as real, but if it is large then they cannot.

This standard error is calculated from the yields of similarly treated plots[1]; it affords an estimate of that

[1] The method of calculation is known as the "analysis of variance".

part of the variation of yield attributable neither to the treatment nor to the position of the plots but to soil inequalities and other disturbing factors such as depredations of pests, inherent differences between individual plants, etc. The wider the ratio between the experimental result and the standard error the more likely is the result to be due to the treatment and not to soil inequalities; the odds are about 20 to 1 if the difference is twice the standard error but about 100 to 1 if it is $2\frac{1}{2}$ times the error. Another way of putting this for those who do not like talking about "odds" is that a comparison will exceed its "standard error" only once in 20 times if there is no real effect of the treatment. For all practical purposes an odds of 20 to 1 is good enough, and so differences greater than twice the standard error are accepted as statistically significant. A report on a modern field experiment should therefore be accompanied by a statement either of its standard error or of its "significant difference", i.e. the difference in yield between treated plots which can safely, or within an odds of 20 to 1, be put down to the treatment and not to other causes.

A valid standard error can be calculated if the position of the treatments within the experimental area is decided by chance and not by deliberate selection, though there may be certain restrictions of arrangement as shown below.

The reason for this "randomisation" is that the standard error measures the variability between similarly treated plots, and can be used for comparing different treatments only if the chance of variation has been the same on all plots.

Several types of replicated and randomised field experiments are now in common use.

RANDOMISED BLOCKS

The plots are arranged in one or more rows, and the treatments are then allotted at random. In the simplest form each block contains all the treatments and is simply a repetition of every other block: there may be three or more of them.

An example is furnished by the eight-block test already mentioned (p. 156) for studying the effects of the plant nutrients nitrogen (N), phosphate (P) and potash (K). The eight plots include these in all single combinations thus:

Control (0), N, P, K, NP, NK, PK, NPK.

The eight plots could be made into one block, but arranged at random, and three or four such blocks, all with random but different arrangements, could be set up, making 24 or 32 plots in all. A great deal of information can be obtained from the results.

(*a*) *The effects of the three fertiliser elements.* Twelve comparisons are possible for each of the three fertilisers: e.g. for nitrogen:

N	compared with	0,
NP	,, ,,	P,
NK	,, ,,	K,
NPK	,, ,,	PK,

each being repeated three times. So for phosphate and for potash. The average of all these comparisons represents the average or "main effect" of the fertiliser.

(*b*) *The effect of each fertiliser on the action of the others.* To find out whether phosphate has increased the effect of the nitrogen, or lowered it, or has been without action, it is only necessary to see if the increase given by nitrogen in the presence of phosphate, namely NP-P, is greater or less than the increase given by nitrogen in the absence of phosphate, namely N-0. This difference is called the "interaction"; it is expressed by the sign $N \times P$ and calculated thus:

$$N \times P = (NP\text{-}P) - (N\text{-}0) = NP - N - P + 0. \qquad (1)$$

Another set of plots also gives the interaction of N and P in presence of K:

$$(NPK\text{-}PK) - (NK\text{-}K) = NPK - NK - PK + K. \qquad (2)$$

Both sets are used for the calculation, the average being taken so that the value actually adopted is

$$N \times P = \tfrac{1}{2}(NPK + NP + K + 0 - NK - PK - N - P), \qquad (3)$$

It should be noted that exactly the same result would be obtained by determining the effect of nitrogen on the response to phosphate. Thus $N \times P$ is the same as $P \times N$. The interactions of N and K, and of K and P, are similarly calculated.

(*c*) *The effect of one fertiliser on the interaction of two others*. The effect of K on the interaction of N and P is obtained by comparing the $N \times P$ effect when K is absent, i.e.

$$NP - N - P + 0,$$

with the effect when K is present, i.e.

$$NPK - NK - PK + K.$$

To calculate it, therefore, we have only to subtract them:

$$NPK - NK - PK + K - (NP - N - P + 0)$$
$$= NPK + K + N + P - NK - PK - NP - 0. \qquad (4)$$

This is called a "three-factor" interaction.

An actual illustration, worked out by H. W. Gardner

of the Hertfordshire Farm Institute, will make this clearer. The crop was potatoes, the plots were in triplicate and the average yields were, in tons per acre:

0	N	P	K	NP	NK	PK	NPK
4·6	6·0	4·1	5·0	6·9	6·6	5·1	8·7

or set out in diagram form:

	1. N	2. NK	3. NP	4. NPK	
Yield	6·0	6·6	6·9	8·7	Total for four plots 28·2
	5. 0	6. K	7. P	8. PK	
Yield	4·6	5·0	4·1	5·1	Total for four plots 18·8
					Difference 9·4

To find the general (direct) effect of N, the plot yields in the lower half of the diagram are subtracted from those directly above, giving the results,

1·4 1·6 2·8 3·6 Average 2·35

The average N effect is 2·35 tons per acre. (The figure could also be obtained by dividing the difference, 9·4, between the above totals by 4.)

Similarly, to get the P effect, the plots on the left half are compared with those on the right, 0 with P, N with NP, K with PK, NK with NPK.

−0·5 0·9 0·1 2·1 Average 0·65

(Alternatively $(3 + 4 + 7 + 8) - (1 + 2 + 5 + 6)$ divided by 4.)

The average K effect is given by comparing 5 with 6, 1 with 2, 7 with 8, 3 with 4.

0·4 0·6 1·0 1·8 Average 0·95

Three questions, the most important in practice, have

now been answered. N has been most effective on the average, K second, P third, and it should be observed that to find these effects the eight plots have been halved in three different ways:

1, 2, 3, 4 *v.* 5, 6, 7, 8,
1, 2, 5, 6 *v.* 3, 4, 7, 8,
1, 3, 5, 7 *v.* 2, 4, 6, 8.

The next question is:

Does the response to N vary with K? On the left of the diagram this question is answered in the absence of P.

$$(6{\cdot}6 + 4{\cdot}6) - (6{\cdot}0 + 5{\cdot}0) = 0{\cdot}2.$$
$$\text{NK} + 0 \qquad \text{N} + \text{K}$$

On the right of the diagram *this same question* is answered in the presence of P:

$$(8{\cdot}7 + 4{\cdot}1) - (6{\cdot}9 + 5{\cdot}1) = 0{\cdot}8.$$
$$\text{NPK} + \text{P} \qquad \text{NP} + \text{PK}$$

The average effect of N and K coming together is therefore 0·5. This is called the NK two-factor interaction.

The NP interaction is measured in the same way:

$$(\text{NP} + 0) - (\text{N} + \text{P}) \qquad (\text{NPK} + \text{K}) - (\text{NK} + \text{PK})$$
$$(6{\cdot}9 + 4{\cdot}6) - (6{\cdot}0 + 4{\cdot}1) \qquad (8{\cdot}7 + 5{\cdot}0) - (6{\cdot}6 + 5{\cdot}1)$$

1·4 2·0

Average 1·7

The PK interaction is as follows:

$$(\text{PK} + 0) - (\text{P} + \text{K}) \qquad (\text{NPK} + \text{N}) - (\text{NP} + \text{NK})$$
$$(5{\cdot}1 + 4{\cdot}6) - (4{\cdot}1 + 5{\cdot}0) \qquad (8{\cdot}7 + 6{\cdot}0) - (6{\cdot}9 + 6{\cdot}6)$$

0·6 1·2

Average 0·9

Is there any advantage in all three, N, P, K, coming together over and above the direct effect and the interactions already measured? This is called the three-factor interaction.

This is answered by

$$(\mathrm{NPK}+\mathrm{N}+\mathrm{P}+\mathrm{K})-(\mathrm{NP}+\mathrm{NK}+\mathrm{PK}+0)$$

or

$$(8{\cdot}7+6{\cdot}0+4{\cdot}1+5{\cdot}0)$$
$$-(6{\cdot}9+6{\cdot}6+5{\cdot}1+4{\cdot}6)=0{\cdot}6.$$

The plots have now been halved in seven different ways and have given the following differences:

Summary

General or average direct effects	Effect of	N	2·35	Each based on average of four effects
	,,	P	0·65	
	,,	K	0·95	
Two-factor interactions	,,	N × K	0·5	Each based on average of two effects
	,,	N × P	1·7	
	,,	P × K	0·9	
Three-factor interaction	,,	N × P × K	0·6	Based on one effect only

As just summarised, the results have not the same degree of reliability: the general effects are based on averages of four figures, the first-order interactions on only two, the second-order on one. If they are to be tested with the same measure ("Standard Error") the first-order interactions must be divided by 2 and the second-order by 4, and when this is done the results are in their final form:

N	K	P	N × K	N × P	P × K	N × P × K
2·35	0·95	0·65	0·25	0·85	0·45	0·15

and the relative importance can be seen at once.

The introduction of the factors $\frac{1}{2}$ and $\frac{1}{4}$ also has the important function of making the effects and interactions additive. The effect of N in the presence of P is now $N + N \times P = 2{\cdot}35 + 0{\cdot}85 = 3{\cdot}20$, etc., the effect of N and P together is $N + P = 3{\cdot}00$, and the effect of all three fertilisers is $N + P + K + N \times P \times K = 4{\cdot}10$. This is of course equal to the difference of $NPK\text{-}0 = 8{\cdot}7 - 4{\cdot}6$.

It should be noted that the joint effects of two or more fertilisers do not involve the two-factor interactions.

The standard error for testing the treatment effects was 0·31 ton per acre and the "significant effect" is 0·62 ton per acre; the effects of all three fertilisers, N, K, and P, are clearly significant, as also is the interaction of N and P, showing that the response to nitrogen is increased by the presence of phosphate. The other effects are smaller and not significant.

In planning an experiment the most important decision to make in the first instance is the amount of information that is desired. For practical purposes it rarely happens that the second-order interactions are of sufficient interest to justify much expenditure of time and trouble, and it can therefore be decided at the outset to discard them. By doing this the design of the experiment can be simplified. Expression (4) (p. 273) is not wanted, and so the four plots NPK, K, N, P need not be mixed with the four NK, PK, NP, 0, but can be kept in separate blocks. The experiment therefore becomes reduced to:

Three blocks containing	NPK,	N,	P,	K
and Three containing	0,	NK,	NP,	PK

Still twenty-four plots, but with blocks of four instead of eight plots each, and each block receiving the same amount of the three fertilisers. The arrangement determined at random might come out as follows:

N	P	NK	PK	P	N
NPK	K	0	NP	NPK	K
0	NP	K	P	NP	0
NK	PK	N	NPK	PK	NK

The halving of the blocks means that the chances of soil variations within each block are diminished; the standard error is therefore probably decreased. It is true that the second-order differences cannot be determined, but it was decided at the outset that they were not wanted, and by discarding them the rest of the information, the direct fertiliser effects and the first-order interactions, is obtained with more accuracy. This method of simplifying an experiment by deliberately mixing up or "confounding" certain effects (such as this particular interaction) with block differences is now much used.

FACTORIAL DESIGNS

In the above experiment all possible combinations of the different factors were used: it is a simple example of a factorial design. Nitrogen at two levels (nil and a single dose) was combined with phosphate at two levels (nil and one dose) and with potash at two levels. The number of treatments was 8, i.e. 2^3. If a fourth factor, e.g. lime, were introduced, also at two levels, nil and a single dose, the number of treatments to maintain the factorial design would become 16, i.e. 2^4.

The eight-plot test is admirably adapted to determine whether the crop and soil respond to nitrogen, phosphate, or potash, but it does not give any information on the amount of fertiliser necessary to produce the response. Half as much might have been as good or worse or even better. To attempt to answer the additional question: "how much of the individual fertilisers are desirable?" at least three levels of application must be used, e.g. none, a single dose, and a double dose. The number of treatments jumps from 2^3 to 3^3, i.e. from 8 to 27, and only one repetition would involve 54 plots.

To get over this difficulty of many plots, which is a very serious one on commercial farms, a design has been worked out in which the second-order interactions are used as the measure of the standard error. Since in these interactions there will be included with the ordinary experimental error a little of manurial effects the test will not err on the side of leniency, and false conclusions are, if anything, less likely to be drawn with this design than where there is direct replication.

In the extensive series of experiments organised by the Rothamsted Staff and conducted for several years under the aegis of the Sugar Beet Research and Education Committee the quantities of fertilisers tested were:

Nitrogen at the three levels	0	0·4	0·8 cwt.	Nitrogen (N) per acre applied in the form of sulphate of ammonia
Phosphate at the three levels	0	0·5	1·0 cwt.	Phosphoric acid (P_2O_5) per acre in the form of superphosphate
Potash at the three levels	0	0·6	1·2 cwt.	Potash (K_2O) per acre in the form of muriate of potash

The levels are denoted by N_0, N_1, N_2, etc. Since all combinations are used the first nine treatments will be

$$\begin{array}{ccc} N_0P_0K_0 & N_0P_1K_0 & N_0P_2K_0 \\ N_0P_0K_1 & N_0P_1K_1 & N_0P_2K_1 \\ N_0P_0K_2 & N_0P_1K_2 & N_0P_2K_2 \end{array}$$

the second nine will be these with N_1 substituted for N_0; the third nine will be these with N_2 substituted for N_0.

These twenty-seven plots could be arranged in a single block but this, with no correction at all for soil variation, might give undesirably high experimental errors. By suitable choice of plots three blocks of nine plots can be selected satisfying the conditions that each block contains three plots receiving N_0, three receiving N_1, three receiving N_2 and the same for P and K; also that all of the two factor interactions occur equally frequently in each block.

For the main comparisons, e.g. N_0 with N_1 and with N_2, nine N_0 plots (three from each block) can be compared with nine N_1, and nine N_2, so that these comparisons are made with considerable accuracy in spite of the absence of true replication. The two-factor interactions are based on comparison between totals of three plots (which is half the number available in an eight-plot test with three replications), and they are, therefore, measured with only moderate accuracy.

There are four possible groupings which satisfy the above conditions, and they can be very simply and compactly expressed in the form of sets of three 3×3 latin squares.

		Grouping I			Grouping II			Grouping III			Grouping IV		
		P_0	P_1	P_2	P_0	P_1	P_2	P_0	P_1	P_2	P_0	P_1	P_2
Block *A*	K_0	N_0	N_1	N_2	N_0	N_1	N_2	N_0	N_2	N_1	N_0	N_2	N_1
	K_1	N_1	N_2	N_0	N_2	N_0	N_1	N_1	N_0	N_2	N_2	N_1	N_0
	K_2	N_2	N_0	N_1	N_1	N_2	N_0	N_2	N_1	N_0	N_1	N_0	N_2
Block *B*	K_0	N_2	N_0	N_1	N_1	N_2	N_0	N_2	N_1	N_0	N_1	N_0	N_2
	K_1	N_0	N_1	N_2	N_0	N_1	N_2	N_0	N_2	N_1	N_0	N_2	N_1
	K_2	N_1	N_2	N_0	N_2	N_0	N_1	N_1	N_0	N_2	N_2	N_1	N_0
Block *C*	K_0	N_1	N_2	N_0	N_2	N_0	N_1	N_1	N_0	N_2	N_2	N_1	N_0
	K_1	N_2	N_0	N_1	N_1	N_2	N_0	N_2	N_1	N_0	N_1	N_0	N_2
	K_2	N_0	N_1	N_2	N_0	N_1	N_2	N_0	N_2	N_1	N_0	N_2	N_1

The position of N_0, N_1, N_2 in the square shows what doses of P and K are combined with it. Block *A* contains plot $N_0P_0K_0$. Block *B* contains $N_0P_0K_1$. Block *C* contains $N_0P_0K_2$.

It can be seen at once from the above that each block contains three N_0 plots, three P_0 plots, three K_0 plots, etc.; that each block receives the same total amount of fertiliser $(9N+9P+9K)$; that one P_0K_0 plot occurs in each block, etc. The blocks differ only in the way in which the three fertilisers come together, i.e. in the three factor interactions.[1]

LATIN SQUARES

This consists of rows of randomised blocks built up into a square like a chess board, but with the restriction that each treatment occurs once in each row of plots and once in each column but not more often. Thus if it is desired to compare the effects of nitrate of soda, nitro-chalk, sulphate of ammonia, and cyanamide, five plots are needed, four receiving the respective fertilisers and the fifth without nitrogen. If we call them Ns, Nc, S, C and 0, write these letters on separate cards, shuffle them and draw them, they might come out in the following order:

C Nc 0 S Ns

That then would be the first row. The cards would be reshuffled and redrawn to make the second row, but with the proviso that if a treatment had already occurred in the same column a further drawing would be necessary. The third and fourth row could be drawn

[1] For an actual example, see *Rothamsted Annual Report*, 1933, p. 175.

in the same way. Thus the final arrangement might come out thus:

C	Nc	0	S	Ns
Ns	S	C	0	Nc
S	0	Ns	Nc	C
Nc	C	S	Ns	0
0	Ns	Nc	C	S

In practice a simpler method is adopted. All "standard" latin squares up to 6×6 have been drawn up[1] and the selection is made by random choice from among them, the rows and columns of the chosen square being then rearranged at random among themselves.

The great advantage of the latin square is that since each treatment occurs in each row and in each column a correction is obtained for soil variation in two directions at right angles to one another; in consequence the standard error per plot is on the average appreciably lower than for randomised blocks.

The drawback to the latin square is that the number of plots required goes up rapidly as the treatments increase. For example, to carry out an eight-plot test in this form would require $8 \times 8 = 64$ plots.

One minor variation of the latin square is often useful. By halving each plot an additional factor can be tested, e.g. two varieties could be tried with the same manurial treatments, or an additional fertiliser could be introduced, e.g. magnesium salts on potatoes. The decision as to which half of the plot is to receive the additional treatment must be made by chance, e.g. by spinning a coin. Comparatively small effects can be measured, since all the plots are being used for testing the additional factor.

[1] R. A. Fisher and F. Yates, *Statistical Tables for Biological, Agricultural and Medical Research*, London, Oliver and Boyd, 1938. 7×7 latin squares have been classified by H. W. Norton, *Annals of Eugenics*, 1939, **9**, 269.

Laying out the plots. The most convenient size of the plots on commercial farms is usually $\frac{1}{60}$ acre. Larger plots are needed if bulky organic manures, e.g. farmyard manure or composts, are being treated, and especially if the experiment is to be repeated for a series of years on the same plots. Smaller plots, down to one rod ($\frac{1}{160}$ acre), can be used for vegetables or other garden crops if the work is well done.

Special care is needed in selecting the site. First of all the land must be carefully inspected: it must not only be fairly level and uniform on the surface but should if possible be fairly uniform in depth, at any rate to about 18 in. To test this, holes 9 in. deep are dug at intervals, and from the bottom of each hole samples of the subsoil are taken with a borer for inspection. The experimental area should be all in one piece, but the blocks of plots need not actually touch, though they should not be separated by more than the width of a path, except to avoid an open furrow, a line of drains or a streak of gravel. The shape of the plots is not very important if the edges that border paths or are likely to be affected by neighbouring plots can be discarded; the plots can form either squares or strips, but strips are often more convenient. In many of the Rothamsted sugar-beet experiments on commercial farms the plots are wide enough to allow two rows to be discarded at each side of the plot so as to eliminate the edge effects and reduce the area to be harvested. As these edge rows are not lifted till after the plots are cleared they serve as useful divisions between one plot and the next. This is not practicable when potatoes are lifted by machine on commercial farms. It is very important to mark the corners of the experimental area permanently so that they can be picked out again without difficulty, thus allowing the plots to be reconstituted in case any of the pegs should be removed. In the Rothamsted experiments the marker is a wooden peg with a yard of stout wire attached; it is sunk in the ground at each of the four corners to a depth below that of the deepest cultivation, but the wire comes up to the surface: in practice it rarely gets torn out.[1] The position in relation to some fixed landmark is also accurately determined, so that if the worst should happen and the wires be lost, or the pegs torn out, it would still be possible to set out the plots once more.

Further details of the new methods, and the methods of working out the results, are given in Technical Communication No. 35 of the Imperial Bureau of Soil Science (F. Yates), Rothamsted Experimental Station, and in "Field Trials: their lay-out and Statistical Analysis", published by the Imperial Bureau of Plant Genetics, Cambridge (J. Wishart). See also E. M. Crowther in *J. Roy. Agric. Soc.* 1936, **97**, 54. R. A. Fisher, *Design of Experiments*, London, Oliver and Boyd, 2nd edition, 1937; this is for advanced students only.

[1] Various other devices are used. Mr Gardner suggests that when the corners of the plots are measured from posts driven into the unploughed border of a field or near a hedge, a shallow hole should be dug around the post and the excavated soil scattered. Posts are often removed but holes are rarely filled up. Also a little sodium chlorate can be scattered around the post to prevent the growth of weeds or grass; this reduces still further the chance of losing the position of the experimental area.

For plots on permanent grass an excellent method is to cut out and carry away a small turf—say 8 in. square—from the corner of every plot. One or two of the holes may get filled up but their position can be readily obtained again by measurement from neighbouring holes.

II. SOME USEFUL DATA

Length and area

For agricultural measurements and field plot work 22 yd. is a convenient unit: it is called a chain:

100 links = 1 chain

100,000 square links or 10 square chains = 1 acre

For ordinary measures:

Length		*Area*	
7·92 inches	= 1 link	144 square inches	= 1 sq. foot
12 inches	= 1 foot	9 square feet	= 1 sq. yard
3 feet	= 1 yard	$30\frac{1}{4}$ square yards	= 1 sq. rod
$5\frac{1}{2}$ yards	= 1 rod, pole or perch	16 square rods	= 1 sq. chain
40 poles	= 1 furlong	160 square rods	= 1 acre
8 furlongs	= 1 mile	or 4840 square yards	
		1 square mile	= 640 acres

The foot is a very ancient measure. It originated in Greece, and was $\frac{2}{3}$ of the cubit, a Chaldean and Egyptian measure, the length of the forearm from the elbow joint to the top of the middle finger. From Greece the foot travelled to Rome and thence to this country. The English foot is a little greater than the old Roman and less than the old Greek. It varied somewhat from time to time but it was standardised in the eighteenth century and again in 1855 when the yard was legally defined as the distance at 62° F. between the centres of two gold plugs in a certain bronze bar deposited in the office of the Exchequer.[1]

[1] For an interesting account of the origin of the English measures, see Sir Richard Glazebrook, *Nature*, July 1931, **128**, 17.

Volume

Imperial measure

2 pints	= 1 quart
4 quarts	= 1 gallon
	= 0·1605 cubic foot
2 gallons	= 1 peck
4 pecks	= 1 bushel
3 bushels	= 1 sack
4 bushels	= 1 sack or coomb
8 bushels	= 1 quarter

The basis is the gallon. Formerly there were several different gallons, but the Imperial, now the standard, was defined as the volume of pure water weighing 10 lb. The American gallon is different; 1·03 go to the Imperial; so for the bushels.

The Winchester quart used in laboratories is about 2 Imperial quarts.

Weights

Avoirdupois		*Troy*	
437½ grains	= 1 oz.	480 grains	= 1 oz.
16 oz.	= 1 lb.	12 oz.	= 1 lb.
	= 7000 grains		= 5760 grains

14 lb.	= 1 stone
112 lb. or 8 stones	= 1 cwt.
20 cwt.	= 1 ton
2240 lb.	

But there are also:

1 butcher's stone	= 8 lb.
1 cwt. in America	= 100 lb.
1 short ton	= 2000 lb.
1 metric ton	= 2204·6 lb.
(1 ton	= 1·120 short tons or 1·016 metric tons)

The pound was a Roman measure, "libra" (hence our "lb."); it was $\frac{1}{125}$ part of the Alexandrian talent =

93·65 lb. It was divided into 12 "unciae" each of 437 grains (our "ounce": this now = 437½ grains). In England there were several pounds, but the Troy pound used for gold and silver and formerly bread remained at 12 oz. and was fixed at 5760 grains, while the avoirdupois used for ordinary goods ("aver" is an old English word for "goods") had 16 oz. and was fixed at 7000 grains. The legal pound is defined as the weight of a piece of platinum marked PS 1844 deposited in the office of the Exchequer.

Metric system

Length

10 millimetres (mm.)	= 1 centimetre (cm.)
10 centimetres	= 1 decimetre
10 decimetres	= 1 metre (m.)
1000 metres	= 1 kilometre (km.)

1 mm.	= 0·03937 in.	1 in.	= 25·4 mm.
1 cm.	= 0·3937 in.	1 ft.	= 30·5 cm.
1 m.	= 39·37 in.	1 yd.	= 91·4 cm.
	= 1·094 yd.	1 mile	= 1609 m.
1 km.	= 1094 yd.	5 miles	= 8 km.
	= 0·62 mile		

Area

1 square metre	= 10·76 sq. ft. = 1·20 sq. yd.		
1 hectare (ha.)	= 10,000 sq. m. = 2·47 acres	1 acre	= 0·405 ha.
1 square kilometre	= 100 ha. = 0·386 sq. miles	1 sq. mile	= 2·6 sq. km.

Weight

10 milligrams (mgm.)	= 1 centigram (cgm.)
10 centigrams	= 1 decigram
10 decigrams	= 1 gram (gm.)
1000 grams	= 1 kilogram (kg.)
100 kilograms	= 1 quintal or doppelzentner (dz.)
1000 kilograms	= 1 metric ton

Weight (continued)

1 mgm.	= 0·0154 grains	1 oz.	= 28·35 gm.
1 gm.	= 0·035 oz.	1 lb.	= 453·6 gm.
1 kg.	= 2·205 lb.	1 cwt.	= 50·8 kgm.
1 dz.	= 1·968 cwt.	1 ton	= 1·016 metric tons
1 metric ton	= 0·984 ton		
	= 1 million grams		

Original basis

1 metre was supposed to be the ten-millionth part of the quadrant of a terrestrial meridian.

1 gram was supposed to be the weight of 1 cubic centimetre of water, at 4° C. (its maximum density). Modern measurements show that neither is absolutely correct, nevertheless the original standards kept in Paris still remain in force.

Volume

1000 cubic centimetres (c.c.) = 1 litre (l.)
100 litres = 1 hectolitre

1 litre = 1·76 pints or 0·22 gallons	1 pint = 0·568 l.
1 hectolitre = 2·75 bushels	1 gallon = 4·55 l.
	1 bushel = 35·4 l. or 0·36 hectolitre

Cross measures

1 cwt. per acre	= 0·37 oz. per sq. ft. = 3·33 oz. per sq. yd.
	= 1·26 dz. per ha.
1 ton per acre	= 1 stone per rod
	= 25·1 dz. per ha.
1 bushel per acre	= 0·90 hectolitre per ha.
1 lb. per acre	= 1·12 kgm. per ha.
1 dz. per ha.	= 0·80 cwt. per ha.
1 kgm. per ha.	= 0·89 lb. per ha.

Some common weights

Water

1 gallon = 10 lb.
= 70,000 grains
1 cubic inch = 252·7 grains
1 cubic foot = 1000 oz. nearly
= 62·4 lb.

(A seventeenth-century standardization: not quite correct.)

Potatoes

1 bushel = About 56 lb. (no legal standard)
= 60 lb. in U.S.A. (legal standard)
1 sack = 112 lb. (legal standard)

Grain: weight of 1 *bushel*

	Commonly measured about	Frequently quoted for sale	Board of Trade Returns	Measured bushels in 1 cwt. (about)
Wheat	62	63	60	1·9
Barley	54	56	50	2·2
Oats	39	42	39	2·7

Conversion factors for fertilisers

To convert	To	Multiply by
Ammonia (NH_3)	Nitrogen (N)	0·82
Nitrogen	Ammonia	1·12
Tricalcic phosphate ($Ca_3(PO_4)_2$)	Phosphoric acid (P_2O_5)	0·46
Phosphoric acid	Tricalcic phosphate	2·18
Sulphate of potash (K_2SO_4)	Potash (K_2O)	0·54
Potash	Sulphate of potash	1·85
Calcium carbonate ($CaCO_3$)	Calcium oxide (CaO)	0·56
Calcium oxide	Calcium carbonate	1·78
Calcium hydroxide ($Ca(OH)_2$)	Calcium oxide	0·76
Calcium oxide	Calcium hydroxide	1·32

Temperature conversions

° F. to ° C. Subtract 32, multiply by 5 and divide by 9.

° C. to ° F. Divide by 5, multiply by 9, add 32.

III. FERTILISER SUBSTANCES CONTAINED IN CROPS, AVERAGE, LB. PER ACRE*

	Dry matter	Total ash	Nitrogen (N)	Phosphoric acid (P_2O_5)	Potash (K_2O)	Lime (CaO)
Wheat:						
Grain 18 cwt.	1630	32	36	15	10	1
Straw 28 cwt.	2650	142	16	7	20	8
Total crop	4280	174	52	22	30	9
Barley:						
Grain 16 cwt.	1400	37	28	13	8	1
Straw 22 cwt.	2080	111	14	5	26	8
Total crop	3480	148	42	18	34	9
Oats:						
Grain 16 cwt.	1540	48	32	12	9	2
Straw 25 cwt.	2350	140	18	6	37	10
Total crop	3890	188	50	18	46	12
Potatoes:						
Tubers 7 tons	3920	148	54	25	89	4
Sugar beet:†						
Roots 9 tons	5050	115	40	16	50	12
Tops 8 tons	2960	254	54	18	90	30
Total crop	8010	369	94	34	140	42
Mangolds:						
Roots 20 tons	5370	387	89	33	203	14
Leaves 8 tons	1650	254	51	17	78	27
Total crop	7020	641	140	50	281	41
Swedes:						
Roots 14 tons	3350	163	70	17	64	20
Leaves 2 tons	705	75	28	5	16	23
Total crop	4055	238	98	22	80	43
Meadow hay:						
1½ tons	2820	203	49	12	51	32

* Based on table in Warington's *Chemistry of the Farm*, pp. 72–4, 1902 edition.

† Analyses from Mentzel und Lengerke's *Landwirtschaftlicher Kalender*, 1930.

IV. SOME USEFUL BOOKS ON SOILS AND MANURES AND THEIR USES

F. Knowles and J. E. Watkin. *A practical course in agricultural chemistry.* 1937. (Macmillan.)

A. D. Hall. *The Soil.* 1920, 3rd edition. (Murray.)

—— *Fertilizers and Manures.* 1929, 3rd edition. (Murray.)

—— *A Pilgrimage of British Farming.* 1913. (Murray.)

N. M. Comber. *An Introduction to the Scientific Study of the Soil.* 1936, 3rd edition. (Arnold.)

G. W. Robinson. *Soils, their origin, constitution and classification.* 1936, 2nd edition. (Murby.)

E. J. Russell. *Soil Conditions and Plant Growth.* 1938, 7th edition, (Longmans.)

—— *The Farm and the Nation.* 1933. (Allan and Unwin.)

INDEX

Printed in the United States
By Bookmasters